THE SCIENCE OF HUMAN DIVERSITY

A History of the Pioneer Fund

Richard Lynn

With a Special Preface by Harry F. Weyher
President, the Pioneer Fund

University Press of America,® Inc.
Lanham · New York · Oxford

Copyright © 2001 by
University Press of America,® Inc.
4720 Boston Way
Lanham, Maryland 20706

12 Hid's Copse Rd.
Cumnor Hill, Oxford OX2 9JJ

Library of Congress Cataloging-in-Publication Data

Lynn, Richard.
The science of human diversity : a history of the Pioneer Fund /
Richard Lynn ; with a special preface by Harry F. Weyher.
p.   cm
Includes indexes.
1. Pioneer Fund (Foundation)—History.
2. Eugenics—Research—United States—History.
3. Intelligence levels. 4. Race. 5. Heredity, Human. I. Title.
HQ755.5.U5 L96 2001   363.9'2'0720973—dc21   2001027579 CIP

ISBN 0-7618-2040-X (cloth : alk. paper)
ISBN 0-7618-2041-8 (pbk. : alk. paper)

OCLC# 467 13162

DEDICATION

This book is dedicated to the memory of:
WICKLIFFE PRESTON DRAPER
1892-1972

Scholar, Soldier, and Philanthropist

Contents

Preface

My Years with the Pioneer Fund

by Harry F. Weyher
President, The Pioneer Fund

On 22 November 1994 ABC's *World News Tonight with Peter Jennings* was replete with somber voices speaking of a small penis being a "sign of superior intelligence," "eradicating inferior people," arresting blacks solely because of skin color, race superiority, and mentally ill Jews. This voice-over was spiced with references to Hitler and scenes of emaciated victims in Nazi death camps.[1]

I watched this broadcast with more than usual interest, because I was president of the foundation which was the subject of the broadcast, the Pioneer Fund. Fearing such tabloid treatment, I had refused repeated invitations from ABC to appear on tape for the program.[2] My fears were justified. What I saw was a grotesque distortion, akin to what one used to see in fun house mirrors.

The ABC broadcast was one of an endless series of attacks on Pioneer and the scientists whom it has funded, dating back almost 50 years, most often by making baseless charges of "Nazism" or "racism," thus sometimes inciting student unrest or faculty reaction. The following also has happened to Pioneer and these scientists: One scientist had to be accompanied by an armed guard on his own campus, as well as guarded in his home. Another scientist was required by the university to teach his classes by closed circuit television, supposedly in order to prevent a riot breaking out in his class. Several scientists had university and other speaking engagements canceled or interrupted by gangs of students or outside toughs. Two scientists asked that all professional communications go to their offices and not their homes since their wives were frightened by the abuse their work engendered. Two scientists who had speaking engagements in Australia needed 50 policemen to rescue them from a mob. At one major university a professor invaded the class of another professor, led a raucous demonstration there, and had to be removed by campus police. The son of one of Pioneer's directors agreed to succeed his father on the Pioneer board, but then withdrew when the son's wife objected, citing social ostracism and physical danger.

This was not all. One state university temporarily barred its scientists from doing any research with grants from Pioneer. Another major university retained a large Boston law firm to investigate Pioneer before allowing its scientists to use Pioneer grants. The TV show "Inside Edition" tried to do an ambush interview of Pioneer's president (this writer) at his law office, and then staked out his apartment, questioning his neighbors at random. Media attacks along the same lines as the Peter Jennings attack were all too common.

How was all of this commotion generated around a tiny foundation whose only activities had consisted of (a) a 1937 study of family size of Air Corps pilots and the giving of some scholarships to the children of those pilots, and (b) hands-off grants for research into human nature at about 60 institutions scattered around the world in eight countries?

This book by Professor Richard Lynn tells the true story of the Pioneer Fund.[3] It needs no introduction, but at his request I will add a few personal observations about some of the main events and about the human side of just a few of the people. What I know firsthand about this history is at odds with the media distortions, which unfortunately constitute the only information that many people have.

My role in all of this began in 1951 when I was a young lawyer. My employer law firm, Cravath, Swaine & Moore, asked me whether I would like to be loaned by them to John M. Harlan for work on a temporary crime commission appointed by Governor Thomas E. Dewey. Harlan, later to become a distinguished justice nominated to the United States Supreme Court by President

Eisenhower (and grandson of a former Supreme Court justice of the same name), was then known as a rising star among Wall Street lawyers, and I was enormously flattered by the opportunity. I spent two years with him and learned that he was indeed a star, a megastar. Later I learned about the Pioneer Fund, and that Harlan was one of its founders.

Then in 1954 I met Wickliffe Preston Draper, another founder of the Pioneer Fund and its chief, although not the only, benefactor. I had completed my work with Harlan and had accepted an offer to join a start-up law firm with two other young ex-Cravath lawyers, where I worked as a corporate and tax lawyer during most of the years recounted here. As luck would have it, Cravath at that time received a query from Draper about retaining a lawyer, and they recommended me. When I met Draper, who was usually addressed as "Colonel Draper," I found him to be highly intelligent, learned, physically impressive, unselfishly patriotic — the same traits I saw in Harlan. We got along famously over the years, and eventually I was handling all of his legal affairs.

In 1958 Draper and Henry R. Guild, a prominent Boston attorney and a director of Pioneer for 26 years, asked me whether I would join the board and become president. By then I knew most of Pioneer's history, and knowing and respecting Draper and Guild and being in awe of Harlan, I immediately accepted.

COLONEL DRAPER

From the mid-1950s until his death in 1971, Draper had me present at all his meetings (not just Pioneer meetings) except rare meetings with a family member or college chum. I also became sole trustee of Draper's inter vivos trust and executor of his will. I like to think that I became his closest and most trusted friend, and I have always tried to be true to that trust. As to Pioneer, I tried to carry on in the way I think would have been wanted not only by Draper, but also by General Frederick Osborn, Justice Harlan and the others who preceded me as Pioneer directors and officers.

Draper gave to many organizations besides Pioneer, and we met on non-Pioneer matters with such people as General Mark Clark and General Troy Middleton (both college presidents at the time), Archibald Roosevelt (the son of Teddy Roosevelt and president of the Boone & Crockett Club), and Peter Scott (son of the British polar explorer and a leading conservationist). On Pioneer matters Draper and I often saw such noted professionals as R. Ruggles Gates (the British human geneticist and botanist), Robert Carter Cook (the demographer), John C. Flanagan (the statistician), and Henry E. Garrett (head of the Department of Psychology at Columbia and a president of the American Psychological Association).[4]

As befits a military man, Draper was a stickler for organization. Every other Thursday at 4 o'clock I went to Draper's apartment, which occupied the top three floors at 322 East 57th Street in Manhattan. Usually these meetings were just for the two of us, and lasted about two hours. Occasionally third persons were there. Often I separately met people at my law office or elsewhere in New York, such as an official of the Metropolitan Museum at the museum, or sometimes out of town (such as Gates in London or Garrett in Charlottesville). Draper attended only one meeting outside his apartment during the entire time I knew him, and that was not with a scientist and had nothing to do with Pioneer's scientific research activities.

Meetings at the apartment followed a fixed routine. The doorman at Draper's building said he had been told to expect you and directed you to an elevator operator, who took you in a small elevator to the penthouse. You stepped off the elevator into a small hallway, which opened into a two storied room running the entire width of the apartment building. Draper stepped forward rapidly with a warm smile and extended his hand.

Draper was tall for his generation, with an erect military bearing and a quick step, seeming to reflect the vigorous life he had led, the army years and the hunting years. He had sparkling blue eyes, appearing slightly owlish behind his horn rims, and his hair was cut short on the sides to mesh with the baldness which dated from his World War I days. His demeanor was formal, and he was prone to use surnames and the word "sir," probably reflecting his family upbringing and later British army training, but he was warm and quick to laugh at incongruities. Although his features would have been called handsome, a more apt term would be distinguished or aristocratic, tending toward the chivalrous.

The big room had an enormous fireplace at the east end, a large oak conference table and four oak chairs in the center, and a 5,000 volume library at the west end complete with rolling ladder to reach the second-story shelves, a library which gave the impression of being well used. Around the walls and on the floor were hunting trophies, including a bongo (a rare African antelope) and all the African "big five" (elephant, rhino, buffalo, lion, and leopard), a gun rack, and an array of classical swords. French doors opened onto balconies on the north, east, and south. A question elicited the answer that on the two floors above were an office, saddle room, card room, handball court, and gymnasium.

Draper led you to the conference table and pointed to a chair across from him. A decanter and glasses were on the table. "Sherry?", he asked, but took none himself.

Then, not more than a couple of minutes after you had entered, Draper said, "Now, sir." There was no wasted time, and he was ready for the business at hand. The extent of his preparation was amazing. Often he would anticipate what you planned to say, and let you know that he understood that point and the conversation could move on. He had a small pad in front of him with only about three lines of pencilled notes - the notes were a bare-bones list of items he wanted to be sure to cover. It was soon apparent that he knew his subject well and needed no notes for the details, that he had extensive background knowledge and an even better memory.

Draper set no limit on the length of the meetings, although he commented that he preferred short ones. He did not seem to tire at all as the meeting progressed but sat erect throughout, even after two hours or more.

When the agenda had been covered, Draper summarized the actions that had been agreed on. Then he again asked, "Sherry?" He accompanied you to the small elevator hall, extended his hand, expressed his thanks, and left you there.

For all his decorum, Draper also had an intense personal side. He had been separated from his boyhood Massachusetts friends for a number of years, first by World War I and then by the years he spent abroad. During this time he went partially bald, with only side fringes of hair remaining. One day he was strolling down a boulevard in Paris when he heard a woman's voice cry "Wick, Wick" (which is how his family and childhood friends addressed him). A chauffeured limousine pulled to the curb beside him. Inside, a boyhood sweetheart - now a beautiful woman - was smiling while holding out her hand to him. The conversation was overshadowed by crisis-level indecision on his part. Ought he remove his hat as a gentleman should, even though it would reveal his newly bald head to her? Or should he do the improper thing — leave his hat on — and keep his secret hidden from the beautiful woman? I know about this because in later years he would laugh and say he would like to relive that moment, and that he had worried throughout the next 30 or more years about the decision he had made. He had kept his hat on!

Draper liked to laugh at his own foibles. Three nurses had attended him after an operation for prostate cancer, and he asked my wife to buy gifts for them and gave her a price range. She purchased three watches from Tiffany, equal in value but each slightly different in design. Although she had them wrapped separately and told Draper to hand one to each nurse when they came in, he decided to simplify things and to let each nurse select her own. He opened the packages and left the three watches on the table in his apartment and told the nurses, when they arrived at the invited time, to go to the table together and make their selections.

"It was nearly a cat fight," he told me. The nurses had begun arguing over the watches and then appealed to him for a final judgment, which is what he had tried to avoid in the first place. "Nurses are fine," he said, "but have only one if possible."

POPULATION INTERESTS

The periods in Draper's earlier life were delineated by St. Mark's School, Harvard, World War I, the Spanish Civil War, long hunting trips around the world, periods in Paris and elsewhere abroad, military training camps, and World War II. His later life was marked by his intensified interest in population matters, especially as they affected the United States.

High on Draper's list of matters needing more study were the nature of intelligence and its mutability (as examined later by Arthur R. Jensen and others), the heritability of personality traits (as later demonstrated by Thomas J. Bouchard, Jr. and others), and the importance of intelligence and other traits on our commerce and industry, and even on our civilization (as later demonstrated by Linda Gottfredson, Robert A. Gordon, and others.)

THE BROWN CASE

Since media coverage of the Pioneer Fund often turns to the issue of race, I am glad that in this book Professor Lynn discusses Draper's views on race. I will take this opportunity to add a few comments.

Racial differences might not have been high among Draper's areas of interest except that the political and social developments of the 1950s and 1960s made the issue salient, especially with some of the untruths put out as "science." Almost at the time I first met Draper, the Supreme Court decided the *Brown* case, holding that the public schools must be desegregated. For some time the case was hardly mentioned by us or by others meeting with us. Professor Henry Garrett, an expert on education, told us integration could be most easily carried out if it began at the first grade and worked its way upward. But he doubted that even this "successful" integration would close the known black-white gap in IQ and achievement test scores. These comments were noted, but seemed not important and were not pursued.

Soon, however, school integration was being rushed forward, with Federal courts even taking over schools. The busing era soon followed. The South was filled with dismay, demonstrations, confusion, and some violence, and these symptoms later spread to some Northern cities. Garrett was saddened by this turn of events, and worried that the public school system would be severely damaged in areas with high black populations. Moreover he predicted a white exodus from those schools.

PROFESSOR SHUEY

The testimony in the four lower court cases which comprised *Brown* included assertions that the black-white intellectual gap would be closed once the new integrated education was made available to blacks. Garrett always said this was most unlikely and, indeed, that the "egalitarian dogma," the belief that blacks and whites are genetically identical in mean cognitive ability, was the "scientific hoax of the century." This dogma, which even then had an almost religious quality, has since become more ingrained despite increased contrary evidence.

Garrett's position received support about this time from three scientists. Ernst van den Haag of New York University showed that Professor Kenneth Clark had misled the Supreme Court as to damage allegedly suffered by blacks from segregation. This was important because Clark's findings were cited by the Supreme Court as demonstrating that segregation was not merely unethical, but harmful — an empirical scientific question. Frank C. J. McGurk of Villanova University showed that the average black-white intelligence gap had not lessened between World War I and the 1950s, despite intervening improvement in the socio-economic status of blacks. R. Travis Osborne of the University of Georgia conducted a longitudinal study of black and white students and showed that the achievement gap widened steadily with age and grade level. Osborne also showed that the heritability of intelligence was about the same for blacks as it was for whites.

Now Garrett introduced a young and previously unknown psychologist who was to change forever the scientific view of the black-white IQ gap. Audrey M. Shuey of Virginia's Randolph-Macon Woman's College, in the days before computerized literature searches, had compiled all the known black-white IQ studies, including not only the massive testing results from the two World Wars but even relevant unpublished theses. She identified 240 such studies to the year 1958, expanded to 380 in her second edition which included tests to 1965. These showed that blacks averaged consistently more than 15 IQ points (about one standard deviation) below whites.

Garrett contacted publishers, including those of his own many successful books, but was unable to find one willing to publish Shuey's tome *The Testing of Negro Intelligence.* They told him it was too "hot." So Garrett approached Draper and Pioneer for funds to pay a printer. Shuey herself found the printer, not a publisher or distributor, and had to store the inventory in her home, from where she filled mail orders personally, one by one. At first only a few orders came in, but word crept from scientist to scientist, and then a few small coupon ads were placed in professional journals. There were several reprintings, as word got around. Later R. Travis Osborne and Frank C. J. McGurk compiled a new Volume II of Shuey, covering all the tests from 1965 to 1980, twice the number of tests as were covered in Volume I. Eventually the book appeared in virtually every major library in the United States and many abroad. It has been undoubtedly one of the most cited books surveying black-white testing.

Draper, who often used military metaphors, spoke to Garrett of the book being "heavy artillery," meaning that the empirical weight of so much testing was overwhelming. At that time only a handful of scientists had any notion that the testing was so widespread and produced such consistent results. Some of the egalitarians tried to discredit the book by seizing on the results of a few tests here and there and arguing that they proved the opposite of Shuey's conclusions, but these attempts never posed any serious challenge to her massive survey. Most opponents avoided trying to combat the book head on, and just ducked it. The late Professor Hans J. Eysenck stated in his 1971 book *Race, Intelligence, and Education*:

> What finally swayed the balance ... was ... Shuey's book ..., which brought together all the

evidence in one convincing volume. ... [N]ever again
could [psychologists] assert with honesty that the
evidence disproved ... genetic determinants in the
causation of racial differences (pp. 20-21).

Later the "Miracle in Milwaukee
Experiment," which began in 1966 under the
supervision of Richard Heber, brought great fanfare
about scientists raising the IQ of black children by
early intervention called "Head Start," thus
seeming to disprove Shuey's conclusion that the IQ
difference was in large part genetic. At that time
Garrett said to me that intervention had already
been tried, and the effect wasn't lasting. "It's like
taking someone's temperature with an oral
thermometer," he said. "If you give them coffee
first, they'll test high. If you give them ice cream,
they'll test low. But in a few minutes they'll test
normal again."

Garrett's analogy was proven all too correct
by later studies, and Heber was eventually put in
jail for embezzling government research funds
from another project. To their discredit the media,
which had been quick to give large headlines to the
initial claims of raising black IQ, was largely silent
on the later discrediting of those claims and the
scandal involving the chief investigator.

Shuey herself lived quietly and never
interacted with any of the major figures in *Brown*.
She was attractive and shy, with a quiet dignity. For
those seeking a hate target, Shuey did not fill the
bill. Anyone assailing her almost certainly would
have cast himself in the villain's role.

HAIRY EAR RIMS; THE FATTENED CALF

Not all Pioneer-aided research produced such tangible results. Professor R. Ruggles Gates, the British author of the book *Human Genetics*, conducted a detailed study of an extended family in India and concluded that the hairy ear rims of some men in the family were linked to genes on the Y-chromosome. At that time most geneticists believed that this chromosome carried no genes except those for maleness. They assailed poor Gates when he claimed otherwise, and name calling quickly ensued.

I had already witnessed the *ad hominem* assault on Garrett for his scientific conclusions about race, but I was astonished to find grown men fighting so viciously over hairy ear rims. Scientific questions, large and small, were being decided by personal abuse, a sort of jousting, rather than in the laboratory. Gates, who had been through a highly charged divorce from Marie Stopes, the birth control advocate, had not lost his fighting spirit, and he responded in kind, but his typical British dignity was reflected in the tone with which he delivered his own strong language. Draper was much amused.

I never saw Gates after the hairy ear rim episode, although once he sent Draper a snapshot of himself atop an Indian elephant, looking very tiny, which he was. He was a delightful companion, a fine scientist and genuinely interested in mankind's welfare.

Professor William Shockley, the Nobel prize winner for coinventing the transistor, was widely scorned by egalitarians. Many I suspect feared Shockley was smarter and better informed than they were and thus able to better them in any debate.

Once Shockley asked for a small Pioneer grant to assist an expert at a California university (not his own) in researching the "identical twin transfusion syndrome," which caused identical twins (who shared the same genes) to be born with different weights and to show different development. The project was race neutral, as was most of Shockley's research. Pioneer agreed, and sent the grant to the university. But the administration apparently decided to freeze out their own expert because he had been selected by Shockley, and allocated the research funds instead to its agricultural school. There a team of veterinarians operated on an unfortunate cow, and rerouted blood flow to two calf embryos to create the desired syndrome and produce one heavy and one light calf. Detailed data were collected and made public, except that nothing was ever disclosed about the cow. Shockley, accustomed to such politically motivated academic roadblocks, merely shook his head and went about his business. Draper used to chuckle over the lengths to which university officials would go to avoid becoming even remotely associated with the race controversy.

JENSENISM

Soon a new dictionary word was coined, "Jensenism." *The Random House Unabridged Dictionary*, 2nd ed. (1993) defined it as "the theory that an individual's IQ is largely due to heredity, including racial heritage." The adjective "Jensenist" and the noun "Jensenite" soon followed.

These new words related to Arthur R. Jensen, who is regarded with something exceeding awe by scientists in the field of human intelligence. Jensen burst onto the public scene in 1969 with a 123-page article in the *Harvard Educational Review* entitled "How Much Can We Boost IQ and Scholastic Achievement?" The article covered several aspects of the IQ debate from the hereditarian prospective, especially the review of twin studies, and it concluded that compensatory education measures (such as the "Miracle in Milwaukee Experiment") have generally failed to have real impact. It also suggested that the average IQ gap between blacks and whites was probably not due entirely to environmental differences. The intensity of the reaction to this last small part of the largest article the *Harvard Educational Review* had ever published was difficult to imagine, and included campus demonstrations, votes of condemnation by various groups, and much media abuse.

I, of course, noted this furor, and I telephoned the *Harvard Educational Review* and asked to purchase a copy. A voice amazingly said that the *Review* had terminated selling any more copies of this particular issue. When I asked why, the voice said that the *Review* was obtaining an "answer" to Jensen, and would publish Jensen's paper and the answer together in a single cover, and that the article would be available no other way. I then asked if I could get permission to reprint the article alone and pay the usual royalty to the *Review*. The voice said that no permission to reprint was being given, even for a royalty.

The *Review* did republish the article together with several commentaries, plus Jensen's reply to them. But none of these rose to the level of a rebuttal of Jensen, so I am still waiting for the "answer" the spokesman promised 30 years ago.

The opposition to Jensen was so intense at one time that he had to be accompanied by an armed guard on campus, and for a while even his home was under guard, and he was advised to have his family move in with friends until things cooled down. It is a sorry example, for which Americans should be ashamed, because no one has ever made any serious challenge to any material part of Jensen's work, or for that matter to even any small part of it. It seemed to me, and it seemed to Draper and others in 1969, that Jensen was headed for greatness. He was, and he has achieved it.

A remarkable trait which I noticed about Jensen over the years is his willingness to help other scientists. He has never shown the slightest evidence of "hogging" anything for himself, even his own work. He feeds out information wherever needed by a fellow scientist, saving many a scientist wasted laboratory time, as well as possible embarrassment. When reviewing an article for publication, a task for which he is in great demand, Jensen will not simply comment, "This article should include a factor analysis (or some other statistical procedure)." If possible, Jensen will perform this analysis and provide it to the author. Where his own area of expertise intersects with another area, Jensen helps the scientist in the other area plan to make the intersection seamless. He has always been as generous as he is great.

The 1969 *HER* article was only a small part of an enormous whole, and a huge flow of original research and scholarly writings has come from Jensen. Jensen has a new 1998 book summarizing much of the work he has done during his lifetime. It is called *The g Factor*. One prominent reviewer called it "magnificent (and awesome)."

Jensen's frustrated opponents have evidenced such a hatred for him, that I am sure many of them secretly wish he would simply vanish from the earth. I told him for that reason it tickled me that the members of his family often live to be over 100. Thus his enemies might look forward with dismay to several more decades of Jensen productivity, probably outlasting most of them.

PROFESSOR SHOCKLEY

Bill Shockley took almost as much undeserved abuse as Jensen. The main charge against Shockley was that he was a physicist trying to be a social scientist, and thus unqualified. But on examination that charge cannot stand. Shockley was a mathematical genius and it was just as easy for him to apply statistical analyses to data collected by social scientists as it was to apply mathematics to physical formulas underlying the transistor. In fact he probably spent a greater portion of his life on human problems than on transistors. He genuinely was concerned with human welfare, both mothers and children, and devoted much of his life to analyzing relevant problems. Shockley was often unfairly called a racist, but his "thinking exercises" and proposals were always in fact race neutral.

Bill Shockley was frequently in New York, sometimes with his wife Emmy. On several occasions he was addressing audiences and wanted to provide typewritten handouts for them. He would use my law office, then at 299 Park Avenue, as a kind of branch workplace. There he would compose and edit handouts, and he and Emmy would work away. I often marveled that we had a Nobel prize winner in his shirt sleeves operating our typewriters and duplicating machines (these were the pre-computer days), with his wife rushing about to help him, often at late hours when our strong, ambitious young associate lawyers had become exhausted and gone home.

Shockley attracted abuse for whatever he did. Possibly this was because his intellectual powers so overwhelmed his opponents that they were reduced to *ad hominem* abuse as a last resort. Jensen was aware of this, and said, "Shockley is a lightning rod."

Shockley received death threats several times. Once he was in New York to debate anthropologist Ashley Montagu on TV, and he received such a threat. I called Vince Gillen, the private eye involved in the General Motors-Ralph Nader set-to, and he was our bodyguard (presumably armed) that day.

Once Shockley asked John Trevor, a Pioneer director, to set up a luncheon meeting with a prominent physician. At lunch Shockley asked questions about welfare mothers. He perceived that the physician was giving answers aimed at political correctness rather than truth. In this spirit, Shockley asked the man to submit himself to a lie detector test. Their acquaintance was short.

But Shockley was also a fun lover and a bit of a cut-up. Once my wife and I invited him to the 21 Club for dinner. Somehow he wandered into the nearby area of Times Square while it was still a "combat zone," before its recent rehabilitation. While he seemed to have escaped the worse dives, he did get into a shop selling magic tricks. When we were seated later on a banquette downstairs at 21, he literally thrilled a bejeweled lady sitting next to him, with all sorts of tricks, using cards, dice, handkerchiefs, etc. I imagine she had gone there to be seen by other socialites, not shown magic tricks by a stranger, and I doubt she knew she had a Nobel prize winner on her hands. But at the end I'm sure she would have said he was worthy of that prize. And, on reflection, perhaps Shockley really was a magician.

On another occasion Shockley said he wanted to have dinner at the elegant Côte Basque, because someone had recommended it. When we were seated Shockley announced he was on a diet, because he had to lose 20 pounds. First he ordered more than one large martini, which must have eaten up a good bit of his calorie allowance. Then he told the waiter he wanted "just a head of lettuce", nothing more. Since the dinner was prix fixe at about $40 (a big price for those days), that must have been the most expensive head of lettuce in town. It arrived on an oversized round platter, laid out flat like a giant pressed flower, with pimento in the center to add color, and lemon slices on the side. Shockley seemed happy with this weight-losing fare, so he ordered some wine to go with it. I can't remember whether he had dessert and/or brandy, but I wouldn't be surprised.

I meant to ask later about the results of his diet, but I never got around to it. I suppose the diet wasn't successful because a short time later on TV's Phil Donohue Show, dispelling any claims that he thought himself a kind of "genetic superman," Shockley acknowledged being overweight and pirouetted before the national TV audience to model his slightly protruding middle.

PROFESSOR EYSENCK

Hans Eysenck's death in 1997 caused, not merely the usual grief, but an outburst of reverence, as well as quiet homage from those who felt honored to have worked with so great a man.

Eysenck was born in Germany, the son of two silent film actors, but resisted the urge to follow them in that career. He left Germany rather than be pressured to join the Nazi party, and he studied in France and later England. At the famed University of London he studied under Sir Cyril Burt. In 1950 he started up a new psychology department at the Institute of Psychiatry and headed it for many years. His contributions to psychology were numerous and important, including his work on the role of genetics in personality and intelligence, and his debunking of psychotherapy and the Freudian overemphasis on environment and child rearing practices.

Eysenck was a soft-spoken and gentle man who would patiently answer a silly question, such as I might ask, as readily as a serious one. Knowing his quiet nature makes it difficult to envision the people who once physically attacked him in a lecture hall (where one of his young students, not otherwise known for physical combativeness, named J. Philippe Rushton, then adorned with the long hair of the times, rushed from his seat to help defend Eysenck) or the mobs which sometimes tried to bar Eysenck from speaking.

When Jensen's famed 1969 paper was published in the *Harvard Educational Review*, suggesting that the black deficit in educability was partly genetic, the roar of the anti-Jensen forces was so loud as to drown out the numerous but less noisy supporters of Jensen — except one. One voice came through on Jensen's side, loud and clear, and that was the voice of Hans Eysenck, the man who years earlier had attracted Jensen to study under him at the Institute of Psychiatry. In 1971 Eysenck wrote *Race, Intelligence and Education*, agreeing with Jensen. From that time on the voices of still other Jensen supporters could be heard getting louder. Some of Eysenck's legion of bright students also took a fresh look at the question and began to make themselves heard. The pendulum was swinging, at least among scientists, partly due to Eysenck.

To scientists and scholars Eysenck is just as alive after his death as before. His 100 scientific books and 1,000 scholarly articles insure that. And scientists and scholars in the future will invoke Eysenck's integrity and brilliance, just as New York Yankee baseball fans invoke the dignity, character and power of their former center fielder to the tune of "Mrs. Robinson":

> Where have you gone, Joe DiMaggio?
> A nation turns its lonely eyes to you ...

and to another tune:

> Jolting Joe DiMaggio
> We want you on our side.

Jolting Joe, of course, died recently but still lives as an inspiration to the Yankee side, which is the right side for those fans. Eysenck is still with us too, intellectually, and he too is on the right side — the side of truth.

PROFESSOR BOUCHARD

Ten years after Jensen's famous article, Thomas Bouchard and his Minnesota team came onto the national scene. Starting in that year and continuing to the present, they have studied over 100 sets of identical twins from several countries who were separated at birth and reared in different environments. These twins are compared with each other, with blood siblings and relatives, and with adoptive siblings and relatives. The broad conclusions: (1) genetic factors have a pronounced influence on social attitudes, vocational interests, psychological characteristics (including IQ) and certain behavior; and (2) the effect of being reared in the same home is negligible for many traits. Bouchard's conclusions accord with Jensen's (and with those of Joseph Horn's University of Texas adoption study, also supported by Pioneer).

Bouchard always rested his case on the scientific analysis and resisted the temptation to place any significant weight on the many interesting anecdotal aspects, such as twins marrying wives with the same name, twins selecting the same breed of dogs, twins giggling constantly at the same things, twins wearing the same number of rings, and the like. Much of this latter material found its way into the popular media, but one tale I heard from Bouchard was never printed. Identical male triplets, although raised separately, had each developed a penchant for pinching the bottoms of uniformed nurses, and then laughing joyously. Bouchard said this could have caused problems when the triplets were reunited in a Minneapolis hospital, where there were many pretty nurses in the halls, except that the nurses looked upon the triplets as laboratory specimens, and viewed the pinching as important scientific data.

Amusingly, although Tom Bouchard has been vilified by radical student groups and professional demonstrators as a reactionary, he himself as a Berkeley student was arrested and jailed overnight for participating in a student takeover of the campus administration building, being bailed out the next morning by his pretty young bride. I don't know when Tom lost most of his hair, as he has, but if he was balding then and since he is quite tall, he must have been a sight to see in the holding pen.

THE OTHER SIDE OF SCIENTISTS

Sometimes a scientist would say he'd like to visit Draper to thank him for the financial support. Draper always told me he'd be glad to see the scientist on a social level, and hear about his activities, but he preferred not to do so unless the scientist persisted. His reason was to avoid the appearance of trying to influence the scientist. In fact he never met with more than a handful of all the scientists whose research was financed by Pioneer in the post World War II years. And I believe that, except for Garrett, Gates, Cook, and Flanagan, he never saw any scientist more than once.

I felt no such constrictions, although I preferred to avoid discussing work. Whenever a scientist was in New York for a seminar or otherwise, I offered to take him to lunch or dinner if our schedules permitted. And a few times I have been to seminars elsewhere.

Tom Bouchard and his wife Pauline, who has degrees in both law and biology, are delightful companions. The same is true of Art Jensen and his psychologist wife Barbara, and Phil Rushton and his anthropologist wife Elizabeth. Others also had wives with careers of their own, such as Sybil Eysenck, who stays so busy she had to decline a nomination for the presidency of the International Society for the Study of Individual Differences. The late Susan Lynn worked with her husband on the worldwide increase in intelligence in the last 60 years, the contributions of improvements in nutrition, and the intelligence of Oriental peoples. Maybe philosopher Mike Levin has the best spousal deal occupation-wise, because his wife Meg also holds a Ph.D. in philosophy and takes over classes for Mike whenever he has to be away.

Henry Garrett was a fun person I saw often. Once I attended an academic convention with him and his wife in the Virgin Islands. Mildred was a small, vivacious, and popular woman with a piercing voice. She could be heard for blocks when she was excited. When we arrived at our hotel, Garrett discovered that he had come from the airport with the wrong bag. The one he now had contained only ladies underthings. He grumbled. Later on, during a day of shopping for items which were cheaper there, he and Mildred bought some scotch whiskey to take home. After the convention, when we were in the departure room at the small airport, Henry was gathering the baggage and still grumbling about the mistake. Just at a quiet moment, Mildred's piercing voice filled the room, "Forget about those undies and bras you had, Henry. Where's the whiskey?" Garrett was the instant center of attention from his fellow academics, who pressed him amusingly but unmercifully to "come clean" and "tell the real truth, Henry."

Ruggles Gates, also a good companion, once took me to lunch at the Army & Navy Club in London, which was an archetypical English men's club. At that time I was a chain cigarette smoker (no longer, of course). Gates was speaking gravely of boorish academics, who did not know how to behave as gentlemen, with which I suppose I was expected to agree. But before I could say anything about such boorishness, three waiters in black suits, shiny black shoes, white shirts and black bow ties, descended on me from different directions and pushed an assortment of plates and bowls under me, all of them trying to help me snuff out the cigarette, informing me with English graveness and propriety that smoking was not permitted! Men at other tables briefly looked at me without expression, unless it was disdain. This was in the late 1950s or early 1960s and the first non-smoking club I'd ever been in. England was years ahead of us. I don't think Gates ever invited me to lunch again. I was boorish, I know.

FRACTIOUS ACADEMICS

The scientists mentioned above, and others I will mention later, are worldwide leaders of current scientific thought in their fields. Their work, together with the work of certain others such as the late Richard Herrnstein of Harvard, who was not a Pioneer grantee, constitutes the scientific mainstream today.

Their agreement on many aspects of the race, genetics, and IQ issues makes them seem monolithic to those with only a cursory knowledge of their research, and this enables their opponents and some of the popular media to portray them as all marching in lockstep. But this they aren't! Each marches to the beat of his or her own research program, and they not infrequently disagree with each other.

Perhaps the most dramatic episode involved William Shockley. He wrote a letter to General Frederick Osborn concerning possible research on racial differences as these might be affected by dysgenic breeding patterns. Osborn, who had been president of Pioneer itself at an earlier time and also president of the American Eugenics Society, had written a definitive pre-war (1940) book on eugenics entitled *Preface to Eugenics*, in which he stated at p. 78:

> It is very important that there should be further scientific studies on the genetic capacities of the different races.

In his revised edition (1951), that sentence was deleted from the corresponding paragraph on page 122. This deletion probably reflected the rising influence of the Franz Boas egalitarian school after World War II, because the 1951 edition adds a citation to Otto Klineberg, an anti-hereditarian and a Boas disciple, and describes him as being an expert on "international psychology," (p. 121).

Now on 28 February 1968 Osborn answered Shockley in a private letter that said:

> so far there is no evidence that white people are on the average (above) Negroes in their genetic potential for intelligence.

He then referred by name to the book *The Testing of Negro Intelligence*, but he apparently did not remember the author's name and referred to Shuey as "some woman," and said:

> I think you will find that every competent sociologist or psychologist who reviewed it [Shuey's book] considered it fatuous to the point of being childish.

Of course time has been kinder to Shuey than Osborn was, but the episode illustrates the professional schisms that can and do occur between knowledgeable people such as Osborn, Shockley, and Shuey, all three linked to Pioneer.

A long time director of Pioneer, John B. Trevor, Jr., has been active in many historical, patriotic, and conservation societies. I once asked him why some of these did not consolidate, since their functions overlapped almost completely. "Don't you see?" he said, "these people are prima donnas, and they want to be leaders, not followers. They'll cooperate but not obey, and I wouldn't have it any other way."

And that's the way with top scientists. They are independent thinkers. These "pie in the face" fights are a normal and healthy expression of that independence, and often lead to clearer delineation of scientific issues.

In fact, I can laugh to myself at the thought of Bill Shockley being locked in a clique with anyone. He was far too strong willed, and brilliant, to go in lockstep with anyone, except his wife Emmy, who seemed to have his number. Others shortly would be exhausted, frustrated, and bewildered, if not worse. Or perhaps it could be said fairly that Shockley was a one man clique.

Knowledge of the relevant research shows how "un-cliquish" Pioneer grantees are. Arthur Jensen and T. Edward Reed disagreed in the journal *Intelligence* with some of fellow grantee Philippe Rushton's work on Asian and white cranial capacities. Rushton argued back in the same journal, and eventually he and Jensen found themselves in agreement, although Reed has not indicated how he stands now.

One Pioneer grantee even wrote a letter saying that the conclusion in a scholarly article by another grantee was "absurd."

Even more interesting is a tale of strange bedfellows. Rushton has done a blistering review of some of Stephen J. Gould's work, which certainly could not have generated any love between them. Prior to that time Rushton began to use the "Out of Africa" model of emerging man in some of his theorizing. Pioneer grantees Seymour Itzkoff and Roger Pearson were less than happy, given that they had endorsed the opposing view of "Multiregionalism," which had been espoused by the famous anthropologist Carleton Coon in the 1960s. Once some of them (to remain unnamed) happened to be in my office in New York, and when I had to step outside for a moment I heard loud voices inside, even rough edged, not quite shouting but nearly that. Later Rushton found his position supported by none other than Stephen J. Gould, of whom he was later so critical. It is hard to think of Rushton and Gould as figurative bedfellows, but this is what happened.

A different kettle of fish is the vicious and unfair *ad hominem* attacks made against some scholars by people whose goal is to conceal or smother honest research, sometimes by demonstrations by political groups such as the Students for a Democratic Society (SDS) in the 1960s, sometimes by slanted reporting in certain papers or broadcasts, and sometimes even by other scientists whose political views overpower scientific ethics. Some of these attacks are mentioned in this Preface. I refer to them here, because Professor Lynn's interest as he wrote this book was focused on the scientific research, and in the book, as in his own personal experience, he does not pay much attention to that sort of distraction. Although he personally has been subjected to much abuse, as a dedicated scientist he has been able to shrug it off as irrelevant to science.

PORTRAIT OF A SWASTIKA PAINTER

Unbeknown to Pioneer for several years, a figure had been lurking on our trail. It was Barry Mehler. He had done his doctoral thesis on the subject of the American Eugenics Society (now the Society for Social Biology), and there had come across the names of the above-mentioned General Frederick Osborn (nephew of Henry Fairfield Osborn of the American Museum of Natural History) and Harry H. Laughlin, two of Pioneer's founders who also had been prominent leaders in the pre-war eugenics movement in the United States.[5] Mehler always unfairly blurred eugenics and genocide as being nearly one and the same,[6] so now he saw Osborn and Laughlin as contaminating Pioneer. He wrote a total of eight articles about Pioneer, but the early ones were published in such obscure places that we did not even hear of them for a couple of years or more.

Mehler has never attacked the scientific integrity of the Pioneer-funded research. He seems not competent for that. Instead, his attacks were *ad hominem*. With disregard for fact and decency, he has repeatedly labeled people as "racist," "Nazi," or "Nazi sympathizer," and he has tried to link Pioneer and its founders, directors, and scientists to the Holocaust, from Osborn and Laughlin in the early days down to Nancy Segal, then at the University of Minnesota.

When we belatedly heard of Mehler's writings and read them, we were startled that an academic figure would have such a ragged and disjointed style.[7] I later asked a writer friend to look at them and then tell me how to describe the author. "He's a Swastika Painter," she said, "because that's his line of business."

"No," another female acquaintance said, "he's a Stalker." I think she chose this description because Mehler has seemed obsessed with Pioneer for all these years and followed its actions from a distance, but never to this day has he sought access to its files or met a single officer or director of Pioneer. I don't believe he has ever met any of the scientists either, except Philippe Rushton briefly on the Phil Donohue and Geraldo Rivera TV shows, and except that he once interrupted a symposium honoring Lloyd Humphreys, head of the psychology department at Illinois, by shouting that Humphreys' research would lead to Nazism and a holocaust.

When we first discovered Mehler trailing us, we were already stuck with him in one sense. His innuendo, quotes out of context, guilt by remote association, proof by tautology, name calling, gross distortions, and the like had already been spread in many places.[8] Copies of his materials had gone mysteriously to numerous campuses, and copies were sent anonymously to many in the media.

Mehler has appeared in person on several campuses to spread his materials. He has spoken in public where possible and has sometimes gotten himself on TV or radio by claiming to expose Nazis. Mehler personally sat with the Peter Jennings staff to plan the TV broadcast described at the beginning of this Preface, and he gulled the staff. The first paragraph of this Preface lists some of Mehler's false and misleading materials, which were accepted in toto by the Jennings staff, eager to provide excitement.

During Mehler's entire career, he has done little of note, other than these attacks on Pioneer, and except for brief similar attacks on the deceased Stanley Porteus of the University of Hawaii and on 92-year-old Professor Raymond Cattell of Hawaii, using the same tactics.[9] The former attack sought to change the name of Porteus Hall, originally named in his honor. The latter attack caused the American Psychological Association to postpone a lifetime achievement award it had planned to give Cattell in August 1997. Then, sadly, Cattell died in February 1998, while still defending himself against Mehler's charges, and before receiving his award, shortly before what would have been Cattell's 93rd birthday.

One scientist said, after hearing one of these false charges, "Mehler has tied a tin can to his own tail, which he'll realize some day — too late!"

Mehler, intentionally or not, has fostered copycats such as Adam Miller, John Sedgwick, and Charles Lane, all tabloid type writers who did not hesitate to adopt Mehler's charges as their own, embellish them, and in some cases add vivid untruths of their own creation. Miller's writings, like Mehler's, have been fed mysteriously to the media.

PAINTED INTO A CORNER?

I was saddened but not surprised when overly eager newspaper and broadcast reporters and magazine writers repeated these Mehler-Miller stories. The stories were easy to find, they were titillating, and they sold papers. The more responsible papers, including the *New York Times* (21 February 1996), the *Economist* (24 January 1998), and the *Wall Street Journal* (9 January 1995 and 22 June 1999), promptly published letters from Pioneer correcting the record. Some others, however, including London's *The Independent on Sunday* and the *Sacramento Bee*, required prodding by Pioneer's lawyers before publishing Pioneer's corrections (8 July 1990 and 9 March 1996, respectively).[10]

I was surprised and dismayed to find that a few academic figures, even highly placed ones, fell into the same trap. I had expected them to be more responsible. The list of academics who repeated the false stories, without contacting Pioneer or its officers or directors and apparently with little or no attempt to verify the facts with non-hearsay evidence, includes Richard Delgado of the University of Colorado, Michael J. Howe of the University of Exeter, William H. Tucker of Rutgers University, Thomas F. Pettigrew of the University of California at Santa Cruz, and Michael Shermer of Occidental College.[11]

One of these, Professor Tucker, after being criticized by me in the journal *Society* for repeating unverified (and false) hearsay, simply found some new unverified (and false) hearsay and substituted that, in the same journal. Some of them refused to acknowledge error, huffing and puffing while being unable to produce any non-hearsay evidence to support their claims. My writer friend, referred to earlier, said we might call this "the trapped rat syndrome," but I suppose that's a little harsh.

Perhaps we should just say they painted themselves into a corner.

This little group of academics is disproportionately influential within many professional organizations. When Pioneer filed a formal complaint against a professor of psychology and against his friend as an editor of the *American Journal of Psychology* for refusing to retract a falsehood, Deborah Carliner of the Office of Ethics of the American Psychological Association replied with a letter to this writer dated 28 July 1997 to the effect that an ethics charge could not be opened against the two men, and that the policy of the Office of Ethics was not to disclose the reason.

Of course there are some academics and others who are motivated by fear or career considerations, rather than political beliefs. I wish Pioneer had a dollar for each time a scientist, a media person or a publisher said, "I pretty much agree with so-and-so or such-and-such, but if I say anything like that, I'm likely to lose my grant/job/promotion/book contract, etc."

Now I realized that Professor Gates' controversy about hairy ear rims had been mere child's play. Mehler-Miller created far more noise, but of course the enormity of their false charges was greater.

BACKHANDED COMPLIMENTS

In the decades of attacks on Pioneer, we could find a little solace in being deemed worthy of attacks, rather than going unnoticed. Also we could take pride in the failure of the attackers to contradict to any real degree the scientific research itself, but rather their fallback on the default tactics of *ad hominem* attacks on scientists or other individuals connected with Pioneer, or Pioneer itself.

Then, once in a while, the attackers paid us a more direct compliment, unexpected by us and certainly unintended by them. The following passage is from Stefancic and Delgado's (1996) *No Mercy* (p. 142):

> It [Pioneer] selects the best proposals from the best scholars and funds them amply; grants of $200,000 and more are not rare. Known as the preeminent sponsor of research in this area, it is spending so lavishly that it appears to be depleting its capital so that it may eventually disappear. But before this happens, it will have achieved a remarkable record. Much of the research relied on in the influential book by Richard Herrnstein and Charles Murray, *The Bell Curve*, for example, was financed by the fund. As the *Chronicle of Higher Education* recently put it, "Whether people revere, revile or review the Pioneer Fund from a safe distance, most say that it has successfully stretched [its] dollars a long way." According to Barry Mehler, a historian who has been studying them for nearly two decades, "The Pioneer Fund has been able to direct its resources like a laser beam."

RESEARCH ON RACE

Most of the scientists mentioned above, except Shuey, did not regard race as his or her main field of research. Many were accused of being racist, but only because of the way in which race fitted itself into their bigger research projects on human nature and human variation. For example, the Gottfredson-Gordon team studied the broad field of job qualification, and on average blacks tested lower for police and firefighter jobs than did Asians and whites, not because Gottfredson and Gordon were researching race but because the tested aptitudes were exactly that.

In addition to Shuey, three scientists, although doing work in other fields, did have periods of their careers when they concentrated on the subject of race. These were P. E. Vernon, Richard Lynn, and J. Philippe Rushton.

P. E. Vernon, a president of the British Psychological Society, compiled the first comprehensive book on Asian intelligence, a work almost equal in scope to Shuey's work on black intelligence, and he showed that Asians score higher on average than Caucasians, just as Shuey showed that the latter score higher than blacks.

Richard Lynn has compiled, more than any other scientist, comprehensive studies of worldwide intelligence patterns, as well as important analyses of the "brain drain," which has lowered the average IQ in Scotland and some other places, the effects worldwide of nutrition improvement, and the like. His latest summary on sub-Saharan Africans, including all known testing through 1995, shows the average black IQ there to be 70, or about a standard deviation below African Americans and two standard deviations below whites.

It has always fascinated me that not a single soul complained about Pioneer financing P. E. Vernon when he showed Pacific Rim Asians to be smarter on average than whites — there hasn't been even a whisper of an objection, not from the media, nor from the Swastika Painter, nor from any academics. But when Jensen, Eysenck, Lynn and others showed whites to be smarter on average than blacks, the skies opened up, and torrents of abuse fell on Pioneer and the scientists. One observer, perhaps in disgust, commented that the difference in response reflected the "peculiar, almost anti-white atmosphere in which racial research must now be carried out."

RUSHTON, RACE RESEARCH, AND HATE CRIME LAWS

Rushton has made racial variation into his primary field of research, at least for the present. In his book *Race, Evolution, and Behavior: A Life History Perspective,* he has delineated to a greater degree than any other researcher the continuum on which many human traits vary by race, especially focusing on brain size and its relation to IQ. Rushton also did testing of highly selected black university students in South Africa and found an average IQ of 85, which would be consistent with a general population average of 70.

Like Jensen, he is always available to other scientists to help them with questions in his areas of expertise. He is much in demand at scientific conferences worldwide, where he exchanges ideas and is busy coordinating his work with that of others. Also like Jensen and the others, he has been the victim of abuse, including unruly mobs running through his college hallway, calls for his dismissal by the provincial premier, investigation by the provincial police, and much more. That he is able to handle his delicate subject so well under the circumstances is a tribute to Rushton's dignity, perception, and intelligence. Indeed his talent extends also to having a good radio voice and a good TV presence. Several have said he has movie star quality.

Rushton is able to face up calmly to even hot-headed critics looking for a fight. Once he was debating the Japanese-Canadian television science commentator David Suzuki before a large audience at the University of Western Ontario. Suzuki spoke, not of the scientific issues, but of racism and the like. Rushton, sticking to science, stated that research with the state-of-the-art technique of magnetic resonance imaging and other methods showed Pacific Rim Asians on average to have larger brains than whites, and whites on average to have larger brains than blacks. There were boos and catcalls. Rushton then said quietly, "If you don't believe me, just get a tape measure tomorrow and go out on the campus and measure heads."

That brought stunned silence, probably because the students sensed that Rushton must be certain of his ground. Rushton's manner was such that his suggestion did not sound racist or even harsh. It came across as a softly stated proposal, with an underlying factual basis.

The suggestion amused me, however, with the vision of a 130 pound student nerd telling a giant student athlete, "I want to measure your head." Only Rushton could have gotten away with suggesting that.

No wonder that the Peter Jennings TV show edited out its interview with Rushton, who was able to stand up to their most highly trained interviewers and fare well. The Jennings staff also edited out Gordon, Gottfredson, Michael Levin, and perhaps others whom they first interviewed on tape for many hours but then ignored for the Pioneer broadcast.[12]

The most frightening part of Rushton's story is how Canada's hate crimes laws were used in an attempt to silence him and spike his research. While in the United States our Founding Fathers gave us the protection of the First Amendment, this is not so in most of the world. In Canada and many Western European nations there are laws against free speech, ostensibly enacted to inhibit "hate" and the spreading of "false news."

The facts in the Rushton case are as follows. As a result of the furor over his 1989 AAAS paper, hate crime laws were turned against Rushton in a series of legal and political battles. The premier of Ontario publicly called for him to be fired, the Ontario Provincial Police mounted a six-month investigation of him threatening him with incarceration, and when that failed to eventuate, the Ontario Human Rights Commission mounted a four-year investigation of him. On another occasion Canada Customs seized a copy of his book, *Race, Evolution, and Behavior*, holding it for nine months while their lawyers read it over to determine if it was "hate literature."

The use of hate crime laws to derail scientific research goes well beyond the race issue. Scholars in the social sciences examining any biological or historical question regarding groups, defined not only by race, but by ethnic identity, sex, age, or sexual orientation are potential targets. Despite the First Amendment and America's tradition of academic freedom, the campaign against Rushton in Canada was used as a jumping off point to attack the academic freedom of Robert Gordon and Linda Gottfredson in the United States and to prevent Pioneer from funding their research. This was so even though Gottfredson and Gordon have never taken a position as to the genetic component in race differences, but only demonstrated their existence and the pragmatic consequences for industry and academia, whatever the cause.

GOTTFREDSON, GORDON, AND ACADEMIC FREEDOM

Linda Gottfredson and Bob Gordon teamed up for some of the most valuable social science research imaginable. American industry widely relies on their work on: (a) the relationship between intelligence and productivity, and (b) the testing of job applicants. It was they who exposed the Labor Department's surreptitious "race norming" and other such tactics. They also exposed the Justice Department Civil Rights Division's efforts to effectively abolish intelligence tests in selecting police and firefighters, raising the horrifying prospect of low intelligence or even functional illiteracy among these public servants. Gottfredson and Gordon have clarified for many of us the role of intelligence in maintaining our standard of living, and even our very civilization.

Yet, at one time abuse was heaped on them — research grants were blocked by the University of Delaware,[13] promotions were denied to Gottfredson and a colleague, credits for Gottfredson's course were restricted, they were treated as pariahs in their own two universities. Yet they persevered.

They persevered, and they won. In every administrative and judicial proceeding at the University of Delaware, Gottfredson emerged the victor, always with Gordon's staunch and resourceful support, and the support of some less timorous members of her own faculty. But it took two years of their time, which could better have been devoted to their work.

Here one can credit Mehler with a short-lived success in what he has chosen to make his life's work, at least thus far. His tracts were circulated at the University of Delaware, as elsewhere, and an English professor there wrote a long letter to the interim president of the University alleging that Pioneer was Nazi or racist, citing Mehler and/or his mentor Jerry Hirsch no less than 28 times. Amusingly, the writer claimed that he had not gotten his (untrue) facts from Mehler. The University's interim president, as many academics might do in today's political climate, chose what he must have thought was the safer road, and assumed Gottfredson, Gordon, and Pioneer to be guilty of something evil, although it was never clear exactly what. But he underestimated Gottfredson, and Gordon as well. Today Gottfredson is widely regarded as a real "star," who has brought international academic prestige to the University, probably more than any other Delaware faculty member ever.

She has carried on her research, written extensively, attended to her faculty duties, testified before a congressional committee, and guest edited a major journal, all as a single mother raising teenage twin daughters. She is someone who won't be forgotten.

PUBLISHING OF RESEARCH

An obstacle faced by many scientists, and one more formidable than student riots or physical threats, has been the media's general unwillingness to publish materials on individual or racial differences in intelligence, even where the materials represent the mainstream of scientific thought. Pioneer witnessed this as early as the 1950s, when publishers refused Shuey's landmark book on Negro intelligence, forcing her to go to private printers and private distribution. Nearly half a century later Arthur Jensen, the world's leading expert on mental ability, submitted his manuscript of *The g-Factor* to three major publishing houses in succession, each of which initially indicated great interest but then without explanation to him lapsed into silence for months, neither accepting nor rejecting the book,[14] until the long silence forced him to go to another publisher and finally to a smaller one independent of the pressures of "political correctness." Scientific journals often have behaved in a similar manner. More recently, the publisher of Philippe Rushton's abridged edition of the scholarly *Race, Evolution, and Behavior* abruptly withdrew the book after 45,000 copies had been printed and distributed, with the publisher claiming it had not known what was in the book.

Snyderman and Rothman, in their 1988 book *The IQ Controversy: The Media and Public Policy*, found that the media leave:

> readers and viewers with the very clear impression that expert opinion is decidedly environmentalist and anti-testing. Our survey of experts demonstrates that this is not the case. The news media have allowed themselves to be influenced by a minority of vocal psychologists and educators whose radical views are consistent with a set of journalistic values emphasizing human equipotentiality and equality of outcome (pp. 233-234).

and so

> have come to see themselves not as deliverers of great scholarship to the world, but as gatekeepers for the politically correct (p. 40).

To combat this gatekeeper syndrome, Pioneer has assisted other nonprofit organizations in printing and distributing scientific articles, monographs, and books, and news about scientists and their research, and also has aided public interest law firms where scientific knowledge might be reflected in legal opinions and writings. These organizations of course are limited in their reach. Moreover these organizations sometimes present a dilemma in that they have some political action agenda (which is quite appropriate given their charter and operating guidelines), while Pioneer has never taken a position on any political issue and does not intend to. All too often, no other means of disseminating vital research has been available outside purely academic journals. These organizations have achieved enough distribution to ensure that a record exists of the more valuable work and is accessible not only to researchers, but to interested members of the public.

Among the organizations that Pioneer has supported for this purpose, usually with only small amounts of money, are the American Immigration Control Foundation (managed by G. Palmer Stacy), Atlantic Legal Foundation (headed by Douglas Foster and Edwin L. Lewis), the Center for Individual Rights (led by Michael Greve and Michael McDonald), FAIR (headed by John H. Tanton, Daniel Stein, and others, and which also does considerable demographic research), the Foundation for Human Understanding (managed by R. Travis Osborne), The Hoover Institution on War, Revolution and Peace (one time grant to help publish a book on population problems), the Institute for the Study of Man which publishes the *Mankind Quarterly* (Roger Pearson), International Association for the Advancement of Ethnology and Eugenics (then managed by A. James Gregor, Robert

E. Kuttner and Donald A. Swan), and the New Century Foundation (headed by Jared Taylor).

In very recent times the previously closed gate to publishing has been partially opened. Rushton wrote a paper on race and sex differences in brain size, which passed peer review at a leading journal, was put into page proof, then suddenly was pulled from the issue and rejected. But Douglas K. Detterman of *Intelligence* stepped forward and accepted the paper, and also asked Rushton to write an editorial about the experience. Somewhat similarly, this writer wrote an article[15] recording Pioneer's funding of scientific research which (because I am not a scientist) was critiqued at my request by several helpful scientists until they thought it in shape for publication. I then sent it successively to two prominent professional journals, each of which rejected it outright, and sent me six reviewers' comments which seemed angry and almost abusive. But Douglas and Carol H. Ammons of *Psychological Reports* accepted it, and sent me the comments of four outside, and one inside, reviewers, all of whom made constructive comments, totally unlike the earlier six reviewers.

Perhaps the gate is opening.

THANKS TO RICHARD LYNN

The foregoing pages reflect some memories about my years with the Pioneer Fund and about some of the scientists and others who've been involved. I owe these people a big debt for permitting me over the years to see not only the marvelous work they have done but also the roadblocks that others sometimes have thrown in their way and how they coped. This close-up view reinforces my belief that truth will win, not only because it is right but also because these exceptional scientists are on its side.

These scientists and I owe a great debt to Professor Lynn for writing this book setting forth so clearly what these scientists have given to mankind. Professor Lynn's book has hastened the day when the media, society-at-large, and certain hostile segments of academia will see through the myriad of distortions and will have the intellectual integrity to recognize openly our debt to these scientists, who include in their number some of the most cited and honored individuals in their respective fields, for their role in reshaping the face of social science and man's knowledge of himself.

New York, NY
January, 2000

NOTES

1. Professor Robert Gordon wrote a 69-page analysis of the broadcast, in which he concluded that the "broadcast was ... slick political propaganda tricked up as news." The Pioneer Fund distributed this letter and a transcript of the broadcast to 462 schools of journalism. The letter may be found at www.pioneerfund.org.

2. Also maligned in the broadcast was *The Bell Curve* by the late Richard Herrnstein and Charles Murray. Like me, Murray refused an invitation to be interviewed on tape for the program.

3. A shorter history was written by me. Harry F. Weyher, "Contributions to the History of Psychology: CXII - Intelligence, Behavior Genetics, and the Pioneer Fund," *Psychological Reports* 82 (1998): 1347-1374. Additional facts are also given in Harry F. Weyher, "The Pioneer Fund, the Behavioral Sciences, and the Media's False Stories," *Intelligence* 26 (1999): 319-336.

4. In all of this I observed one of Draper's characteristics that helped me evaluate later questionable claims of statements he supposedly made and conversations he supposedly had. To my knowledge he never made research proposals to any people below the top. Aside from the prominent people listed above in the text, for the early non-Pioneer project on race crossing in Jamaica, Draper went directly to Charles Davenport, the head of the Carnegie Institute of Washington. In establishing prizes for demographic papers, he went to Davenport and to Harry L. Laughlin, head of the Cold Spring Harbor Eugenics Records Office. In considering projects involving intelligence, he went to Henry Garrett, president of the American Psychological Association. Draper of course knew, and was always on good terms with, people of lesser rank, but when a project was involved, he dealt only with the top. In the same vein, when Draper established Pioneer, he invited only outstanding men to the board: General Frederick Osborn, later a member of the U. N. Atomic Energy Commission and then President of the American Eugenics Society; Harry L.

Laughlin; John M. Harlan, described above; and Malcolm Donald, a prominent Boston lawyer and later a civilian in the wartime Pentagon with a rank approximately equivalent to brigadier general. The pattern extended as well to his other philanthropies and to his business dealings.

5. Teddy Roosevelt, Alexander Graham Bell, John Harvey Kellogg, and countless other prominent Americans were supporters of positive eugenics (helping to increase the number of healthy and bright offspring) long before Hitler came to power. Roger Pearson, *Heredity and Humanity: Race, Eugenics and Modern Science* (Washington, D.C.: Scott-Townsend Publishers, 1996). Indeed, Germany itself enacted some programs of negative or reform eugenics (helping to decrease the number of defective offspring) before Hitler's rise, these being patterned after laws which existed earlier in more than half the United States. To credit eugenics to Hitler or to equate it with the Holocaust is both unfair and inaccurate.

6. Both words are partly based on the Greek *genos*, meaning either: (a) birth or (b) kind or race. "Eugenics" uses *genos* in the first sense, and adds the Greek *eu* meaning well. "Genocide" uses *genos* in the second sense of kind, or race, and adds the Latin *cida* meaning killing. So one word means "well born" and the other "killing a race." The two terms have no necessary connection. Mehler could just as easily have blurred "circumscription" with "circumcision," or perhaps adopted William Safire's example of a malapropism, turning "Tannenbaum" into "atom bomb."

7. His doctoral thesis is in sharp contrast to his later writings. The thesis generally is restrained, documented, and accurate. All the later writings about Pioneer were the opposite in all respects, unfit for any scholarly journal.

8. An amusing example is Mehler's typewritten manuscript, headed "An Edited Manuscript - Final Version Scheduled for Publication in *Patterns of Prejudice* (#419)," which was dated 30 May 1989 and circulated on the campus at the University of Delaware. In this manuscript Mehler discusses whether Professor Philippe Rushton can be called a "racist," and for

proof he says that *Webster's Third New International Dictionary* (G. & C. Merriam Co. 1971) defines racism as:

> the assumption that psychocultural traits and capacities are determined by biological race and that races differ decisively from one another.

In fact, the full definition in that dictionary reads as follows:

> the assumption that psychocultural traits and capacities are determined by biological race and that races differ decisively from one another which is usually coupled with a belief in the inherent superiority of a particular race and its right to domination over others.

Pioneer called attention to this misleading partial quote in the publicly circulated manuscript distributed on the campus, and that partial quote was eliminated in the later published version.

9. The attack on Cattell is described by Glayde Whitney, "Raymond B. Cattell and the Fourth Inquisition," *The Mankind Quarterly* 38 (Fall/Winter 1997): 99-125. For Mehler's background, see Roger Pearson, *Race, Intelligence and Bias in Academe*, 2nd ed. (Washington, D.C.: Scott-Townsend Publishers, 1997), 258-280. Another Mehler incident not recorded elsewhere involved a man who had collaborated with Mehler on several occasions and who falsely posed as a reporter for the *Baltimore Sun*, or sometimes the *Texas Observer*, calling himself "Ben O'Brien." The *Baltimore Sun* confirmed by a private letter to Roger Pearson of 23 February 1988 that it had no such reporter. In 1988 the impostor, and also a man who identified himself as Mehler, telephoned the widow of the recently deceased Professor Robert Kuttner, a Pioneer grantee. Although Mehler had published innuendo against Kuttner, he and "Ben O'Brien" now posed as his admirers, and asked the

widow to give them all of Kuttner's papers. She was suspicious, checked with friends, and then refused.

10. The most commonly repeated false charges of Mehler and Miller are listed on Pioneer's Web site, together with the relevant truth, at www.pioneerfund.org.

11. Professor Richard Delgado is a law professor and presumably trained to be alert for false hearsay. Professor Michael Shermer is the editor of a magazine called *Skeptic*, which prides itself on questioning assumptions.

12. Robert Gordon, Linda Gottfredson, Michael Levin, Philippe Rushton, and perhaps others each spent several hours being taped by Jennings' staff, but all this was reduced in the broadcast to a shot of Gordon a few seconds long. What these scientists said on tape, and which was completely edited out, contradicted the whole thrust of the Jennings broadcast. The Jennings staff had the truth before them, but chose not to use it.

13. Private letter to Harry F. Weyher from Andrew B. Kirkpatrick, the Chairman of the Board of Trustees at the University of Delaware, dated 2 July 1990, said the following:

> No matter whether [racism] is in fact the orientation of the Pioneer Fund or not, that is perceived as the orientation of the Fund by at least a material number of our faculty, staff and students. Without judging the merits of this perception, the board's objective of increasing minority presence at the University could ... be hampered if the University chose to seek funds from the Pioneer Fund at this time.

14. Probably not one publisher wanted to be known as rejecting a book by such a prominent scientist, so each just lapsed into silence. Kevin Lamb contacted one of these publishers and was told "Chances are" the silence was a "a very deliberate decision." Kevin Lamb, "IQ and PC," *The National Review*, 27 January 1997, 40.

15. See footnote 3.

PART I:
The Background and Early Years of the Pioneer Fund

Wickliffe Preston Draper
(Photo taken in American Army uniform after
resignation from British Army in World War I)

Chapter 1

Wickliffe Preston Draper and the Founding of the Pioneer Fund

In 1937 an American millionaire named Wickliffe Preston Draper set up a foundation for the promotion of research into the contribution of heredity to human diversity. Draper called his foundation the Pioneer Fund in honor of the early pioneers who settled and built the United States. This book tells the story of the Pioneer Fund and of the work of the scientists and scholars it has supported during the first 60 years of its existence.

Wickliffe Preston Draper was born in 1891 into a wealthy, upper class New England family in Milford, Massachusetts. The Draper family fortune had been made by his grandfather, George Draper, who had set up the Draper Corporation in Hopedale, Massachusetts, and built it up to become the leading American manufacturer of textile machinery. He had three sons: George, William, and Eben. George, the father of Wickliffe, went into the family business and worked for many years as the Draper Corporation treasurer. William fought in the American Civil War as a brigadier general, served as representative for Massachusetts in the U.S. Congress and was later the U.S. ambassador to Italy. Eben was for a time governor of Massachusetts. Wickliffe Draper's mother was Jessie Fremont Preston. She came from Lexington, Kentucky and was the daughter of William Preston, who served as a general in the Confederate forces in the Civil War, as representative for Kentucky in Congress and as U.S. ambassador to Spain.

Wickliffe Preston Draper was educated at Harvard from which he graduated cum laude in 1913. Later that year he traveled on horseback through Mexico witnessing the insurrections led by Pancho Villa and Emiliano Zapata. On the outbreak of World War I in 1914, Draper joined the British army and was commissioned in the Royal Field Artillery. In 1915 he was posted to the Western Front and fought in the battles of Neuve-Chapelle and Aubers Ridge. A letter written to his parents from France describes some of his experiences:

My dear Father and Mother,

We have had a heavy engagement which you may have read of in the newspapers. On the whole the result has been satisfactory but not decisive.

I have successively been with the wagonline, the battery and assisting in observation in a captured German trench so that I have seen most phases of the action. At no time was I in

special danger but as I was for three days subjected
to heavy shell and rifle fire I had a chance to note
its psychological effects. In so far as I can
generalize, one is first apprehensive, then excited,
next indifferent, and finally extremely tired. From
the guns one sees an empty landscape noticeable for
a frequent flash and puff of smoke in the air which
indicates shrapnel, or occasionally a fountain of
dirt which is the strike of a high explosive. The
air is alive with the whistling and droning of
shells, punctuated with the crash of guns and the
boom of shells exploding. As however the Germans
are compelled to fire largely by guess work and as
our line is thousands of yards in depth, artillery
casualties are relatively surprisingly small. The
unburied dead, which have been quite thick, are a
repulsive feature, but one becomes hardened very
quickly.

Always remember that no news is good news
and that if anything should happen to me you
would be promptly informed of it.

Love to you both and to Helen,

WICKLIFFE

In 1916 Draper was transferred to the Twenty-
Seventh division for a period of service in Greece.
In 1917 he returned to Belgium with the field
artillery. He was present at the battles of Messines
and Ypres, where he was severely wounded. For his
service in the British army he was awarded the
1914-15 British Star Medal and the Belgian Croix de
Guerre. When the United States entered the war,
Draper resigned from the British army to join the
American army. He was appointed a regimental
adjutant at Camp Meade, became an artillery

instructor at Fort Sill and, finally, as the war drew to a close, was promoted to the rank of captain and placed in command of an officer training battalion at Camp Taylor.

At the end of World War I Wickliffe Draper had sufficient wealth and private income for it to be unnecessary for him to work, and he lived the life of a gentleman scholar. He served in the United States army reserve as a lieutenant colonel in the cavalry and from time to time went on training courses including a rifle course at Fort Benning in 1924, a course in military intelligence at the United States Army War College in 1932, and the field officers' course at the Cavalry School in 1937. In the 1920s Draper became interested in archaeology and anthropology. This led him in 1927 to join the French Augeras Mission to the southern Sahara. The expedition discovered a prehistoric human settlement at Asselar, some 400 kilometers north of Timbuktu, including the remains of "Asselar Man." Following the success of the expedition, the French Societé de Geographie awarded him the 1928 Gold Medal, and in Britain he was elected a fellow of the Royal Geographical Society.

In the 1920s and 1930s, Draper spent much of his time traveling. He went on expeditions to remote parts of South America, India, Africa, and China, and spent some of his time hunting. He learned to fly light aircraft. In 1937 he went to Spain as an unpaid newspaper correspondent to observe first hand the Spanish Civil War, and he saw a number of battles, including the fall of Bilbao. When the United States entered World War II,

Draper, now 51 years old, was again called to active service and was appointed senior U.S. observer attached to the British headquarters in India. Here he witnessed the fighting on the northwest frontier between the British and the Pathan tribesmen of Waziristan. On his return to the U.S., Draper's last military appointment was as director of internal security of the Northwest Service Command, with responsibility for the Alcan Highway. From the end of World War II until his death from cancer in 1972, Draper resided in New York City.

WICKLIFFE DRAPER, THE MAN

A thumbnail sketch of the personality and interests of Wickliffe Draper has been provided by Harry F. Weyher, who knew him well as his lawyer, executor of his will, and as president of the Pioneer Fund from 1958:

> I know that until his death few people knew who Colonel Draper was, because he refused public credit for almost everything he did. He refused an honorary degree from a French university for work that he did on early man, and rebuffed suggestions of honorary degrees from two American universities to which he had made substantial contributions.
>
> He learned to his dismay, at one point, of a proposal at a mid-western university to name a departmental library for him, and he went to lengths to have his name removed from consideration. The same was the case with credits in a book on archaeology that he had helped write. The only instances when Colonel Draper, during his lifetime, assented to public credit were his

acceptance of wartime decorations and his acceptance of a French government decoration for being co-head of the archaeological expedition into the Sahara Desert that discovered the remains of Asselar man.

Wickliffe Preston Draper was a Yankee, born in Massachusetts of a wealthy family with roots in Massachusetts and Kentucky. Both his paternal and maternal families produced more than their shares of governors, ambassadors, and high officials. He was even distant kin to three American presidents. Colonel Draper led an exciting and almost unbelievable life. He was present at combat in three major wars. He did big game hunting, always with total regard for the rules of fair chase. He had trophies from five continents, having spent several years in Africa, sometimes hunting completely alone even elephants - without a guide or gun bearer. He was an expert horseman, marksman, swordsman, and swimmer, and even could fly the primitive airplanes of the 1920s. He had a great knowledge of history and more than average knowledge of art and literature.

For many years, in addition to his gifts to the Pioneer Fund, he anonymously gave away all his income, and more, directly to other nonprofit institutions for such purposes as education, art, military matters, conservation, and health.

He was a wealthy man, but he was also a gifted man, and, had his duty not taken him to the life he led, he could have become a successful man in many other areas.

His primary interest, however, was in those population problems that might affect the quality of human life in the future. He was an expert on this subject, for he was a highly intelligent man who studied the subject hard throughout his life, and put himself in the hands of paid tutors to learn the advanced aspects of statistics, genetics, and psychology, as they related to studies of mankind.

As to population quality, he concluded, based on his own travels, dealings with people, and research, that the level of civilization in this country, and indeed its greatness, was made possible by its having a reasonably homogeneous population with a combination of four qualities. These included, of course, physical fitness, and three others that are more difficult to identify - high intelligence, courage, and character.

He recognized that these four qualities can be found in all populations, but equally he recognized that the frequencies differed between nations and between groups. He hoped that man would do nothing to lessen the presence of these qualities in the American population and, conversely, would work to increase them.[1]

DRAPER'S SUPPORT FOR CHARLES DAVENPORT

In the 1920s Draper became interested in human intelligence, and the part played by genetics in individual and group differences. He also developed an interest in eugenics. The concept of eugenics was formulated by Sir Francis Galton in England in the closing decades of the 19th century. The principle of eugenics held that human intelligence and personality are significantly determined by heredity and that therefore it should be possible to improve these qualities by encouraging those who are well endowed with them to have more children, and by encouraging those poorly endowed with them to have fewer. At this time many leading geneticists, biologists, and social scientists subscribed to these views, including Charles Davenport, Harry Laughlin, and Hermann Muller in the United States, and Sir Ronald Fisher and Sir Julian Huxley in England.

As his interests in these questions grew, Draper made contact with several of the leading biological and social scientists of the time who worked in these areas. These included Henry E. Garrett, a psychology professor at Columbia University and expert on intelligence testing who later became a director of the Pioneer Fund, R. Ruggles Gates, a British geneticist and eugenicist and first husband of the birth control campaigner, Marie Stopes, and Charles Davenport, one of the foremost geneticists and eugenicists of the first half of the century. It was with Davenport that Draper established the closest relationship in the 1920s.

Davenport was born in 1866 in Stamford, Connecticut, studied zoology at Harvard, and in 1899 took up a position as assistant professor at the University of Chicago. His interests developed in human genetics and eugenics, and he went to England to meet and discuss these questions with Sir Francis Galton and Karl Pearson.

In 1904 Davenport obtained funds from the Carnegie Institution in Washington to establish the Laboratory of Human Evolution at Cold Spring Harbor on Long Island, of which he was the director. In 1910 he obtained further funds to set up the Eugenics Record Office, also at Cold Spring Harbor, from the Carnegie Institution and also from Mrs. E. H. Harriman, widow of the railroad magnate. He appointed Harry Laughlin as superintendent.

Davenport's principal research interest was in finding people with disorders such as alcoholism, epilepsy, mental illness, mental retardation, and criminal tendency, and in assembling their family pedigrees in the hope of detecting the mode of inheritance. He believed, following Mendel, that these disorders were due to single gene defects. This belief turned out to be largely mistaken, and later in the century it became clear that these disorders are mainly caused by a number of genes interacting with the environment. However, he did have one notable success which was his discovery that Huntington's chorea, the severe physical and mental deterioration typically appearing in early middle age, is inherited through a single dominant gene. The effect of this is that if someone with the gene mates with someone without it, an average of half the children inherit the gene and the disorder.

Davenport was also interested in the questions of race differences and the effects of racial crossing.[2] He discussed these issues with Draper and proposed a study to be carried out in Jamaica. The design of the study was formulated, and Draper agreed to finance it. The investigation consisted of obtaining samples of approximately 80 adult whites, blacks, and individuals of mixed-race (at that time termed mulattos), and the same numbers of children. The samples were matched for social environment and were primarily engaged in agriculture. A large number of tests were given, including approximately 30 anthropometric tests of such characteristics as breadth of nose, length of limbs, dimensions of the head, hair form, skin color, and so forth. The results were that there were significant black-white differences on these measures and great variability among mulattos.

A number of mental tests were also given. Simple musical aptitudes of time and rhythm discrimination measured by the Seashore tests showed that blacks performed better than whites. A number of intelligence tests were administered, including the verbal Army Alpha and some performance tests. The whites averaged highest on all the latter tests. On the performance tests the mulattos scored intermediate and the blacks lowest, while on the Army Alpha the blacks scored intermediate and the mulattos lowest. The mean scores on the tests were given but not the IQs or standard deviations, so it is impossible to calculate the differences between the three groups in terms of conventional IQs. Davenport reported the results in a paper and in detail in a book co-authored with Morris Steggerda, who had carried out the testing in Jamaica.[3]

THE EUGENICS RESEARCH ASSOCIATION PRIZES

Following his initial funding of the Jamaican study, Wickliffe Draper began to meet frequently with Davenport and his colleague Harry Laughlin to discuss genetics and eugenics. It was as a result of these discussions that from 1928 onwards he provided funds to the Eugenics Research Association run by Davenport at Cold Spring

Harbor, New York, for the award of annual prizes for research monographs on these issues. The first of these prizes was awarded in 1929 to J. Sanders, professor of medicine at the University of Amsterdam, Holland, for a monograph on fertility in Europe.[4] It presented statistical data for a number of European nations including Austria, Denmark, France, England, Germany, Italy, Ireland, Scotland, Spain, and Switzerland. The statistics showed both generally declining birth rates in Europe from the middle of the 19th century to the 1930s and dysgenic fertility (i.e., the inverse relation between socioeconomic status and numbers of children that Galton had identified in England in his *Hereditary Genius*).

The prize for 1930 was awarded to Roderich von Ungern-Sternberg for his book *The Causes of the Decline in the Birth Rate within the European Sphere of Civilization*.[5] The book was an exhaustive investigation of the factors responsible for declining birth rates. These included later age of marriage, the decline of child mortality, the costs of raising children, and the increasing use of modern methods of contraception. His overall conclusion was that couples were increasingly torn by a conflict between maintaining their standard of living and the expenses of children and were opting to limit their family size in order to maintain their living standards. This was essentially the economic theory of fertility, which was later advanced by the economist Gary Becker.[6] It views children as consumer goods in competition with alternative preferences. In 1935 the Draper prize went to Serge Androp, a

physician in Gallipolis, Ohio, for his family
pedigree study of mental disorder.[7] Androp
assembled the pedigrees of over 200 mentally ill
patients for three generations and showed a higher
incidence of mental illness among the relatives of
the mentally ill than in the general population,
which he interpreted as indicating a genetic basis
for mental disorders.

A further monograph was Ernest Kulka's
Causes of the Fall in the Birth Rate.[8] Kulka was a
gynecologist at the University of Vienna. His
monograph presented statistics on fertility rates
worldwide, and demonstrated the decline of
fertility in Europe and North America as compared
with the rest of the world and the consequent
diminishing proportion of Europeans in the world
population. He discussed the causes of the decline
in fertility in Europe and North America and
reached similar conclusions to those of von
Ungern-Sternberg. Another of the Wickliffe Draper
prize monographs was Wagner Manslau's treatise
on fertility differentials between social classes in a
number of European countries, North America,
and South Africa and the reasons for these.[9]

THE FOUNDATION OF THE PIONEER FUND

In the early and mid-1930s Wickliffe Draper
became increasingly interested in eugenics issues,
including the genetic basis of intelligence and
personality, dysgenic fertility, and the possibility of
finding measures to improve the genetic quality of
the population. To promote these objectives he
decided to set up and endow a foundation. He

named it the Pioneer Fund in memory of the early settlers in America.

The Pioneer Fund was incorporated on 11 March 1937. The certificate of incorporation stated that the Fund had two objectives. These were, first, to provide financial assistance to the parents of children likely to become socially valuable citizens who would make important contributions to their society. The financial assistance would help the parents educate their children and thereby encourage them to have more children. Children were to be identified from the demonstrated qualities of their parents, based on the assumption that these were inherited. The recipients were to be selected predominantly from settlers of the original 13 states. The second objective of the Fund was to provide grants for research into the study of human nature, heredity, and eugenics. The history of this research will be described in detail in the following chapters. Appendix A provides a full statement of the Fund's original objectives.

DRAPER'S VIEWS ON POPULATION AND IMMIGRATION

Draper set out his views on population in a memorandum written in 1960.[10] In these he stated that he thought it desirable that the future population of the United States be in accord with the values of its founders. By this, Draper meant that the population should not increase too greatly in size, that it should be physically and mentally sound and that it should retain its predominantly Western cultural heritage.

His memorandum serves as a guideline for the directors or trustees of the Pioneer Fund, summarizing his views. It begins as follows:

> This memorandum is to set forth the purposes for which I hope any bequest will be used which is made to the Pioneer Fund under Paragraph (d), Article SIXTH of my last Will and Testament, made by me on June 1, 1960, or any similar bequest to the Pioneer Fund, made under any future Last Will and Testament or codicil hereafter made by me, if I die a resident of New York.
>
> I believe that in a few centuries our country and our planet will, at the present rate of increase, be crowded to the bursting point. Accordingly, a selection between human stocks must occur whether left to chance or planned by man. I believe in planning, and I hope that the above-described bequest to the Pioneer Fund will be used to encourage an increased birth rate in, and to otherwise aid in the education and/or support of, a small group which is deemed to have such qualities and traits as to make the group of unusual value as members of our civilization and as citizens of this nation. I recognize that this would involve some quantitative increase in the population, but I believe the increase would be negligible whereas the qualitative improvement might be large...[11]

Draper believed that immigrants from northwestern Europe would be most easily assimilated into the United States. He therefore liked the national quota system which had been in effect since the 1924 Immigration Act. In the 1950s and early 1960s this policy was challenged, notably by John F. Kennedy in his book *A Nation of Immigrants*, in which he argued that since all

Americans were immigrants or descendants of immigrants, immigrants from all countries should be equally welcomed into the United States.[12] Draper disagreed with Kennedy's point of view and in the early 1960s he provided funds for two committees to commission research on these issues. The first of these was known as the Walter Committee. It was chaired by Representative Francis Walter, co-author of the 1952 McCarran-Walter Immigration Law, which retained the policies of the 1924 Immigration Act. The second was the Eastland Committee, chaired by Senator James Eastland, whose members included Henry Garrett, professor of psychology at Columbia and a director of the Pioneer Fund from 1972-73, and Joseph W. Broullette of Louisiana State University.

DRAPER'S VIEWS ON RACE

Draper was concerned about the social status of blacks in the United States and their relative educational and occupational attainments. He knew of Henry Garrett's view that, on average, intelligence scores were lower for blacks than for whites, and he believed that this subject deserved further inquiry. In the late 1950s Professor Audrey Shuey summarized the published scientific literature on black-white differences in intelligence. When Shuey was unable to find a commercial publisher, Draper provided the funds for private publication of her book *The Testing of Negro Intelligence*, now a standard reference.[13]

DRAPER'S LEGACY

After the end of World War II and until his death in 1972, Draper spent most of his time in New York City, where he dealt with projects in his many areas of interest which also included military history, art, and conservation. In his will, Wickliffe P. Draper bequeathed approximately $5 million to the Pioneer Fund. This bequest has been instrumental in furthering our understanding of the nature of human intelligence and personality, the nature of race differences, and the effects of differential fertility in modern technological societies and of immigration on the fabric of American society. As we shall see in this book, the Pioneer Fund has been able to keep scientific research into these vital questions alive at times when major foundations and public funding agencies preferred to avoid them.

NOTES

1. Weyher, H. F. 1994. Address to private audience. *Pioneer Fund Archives*.

2. Davenport, C. D. 1928. Race crossing in Jamaica. *Scientific Monthly*. 27: 225-238.

3. Davenport, C. D. & Steggerda, M. 1929. *Race Crossing in Jamaica*. Washington, D.C.: Carnegie Institution.

4. Sanders, J. 1929. *Comparative Birth Rate Movements among European Nations*. Cold Spring Harbor, NY: Eugenics Research Association.

5. von Ungern-Sternberg, R. 1931. *The Causes of the Decline in the Birth Rate within the European Sphere of Civilization*. Cold Spring Harbor, NY: Eugenics Research Association.

6. Becker, G. S. 1981. *A Treatise on the Family*. Cambridge, MA: Harvard University Press.

7. Androp, S. M. D. 1935. *The Probability of Commitment for a Mental Disorder based on the Individual's Family History*. Cold Spring Harbor, NY: Eugenics Research Association.

8. Kulka, E. 1931. *Causes of the Fall in the Birth Rate*. Cold Spring Harbor, NY: Eugenics Research Association.

9. Manslau, W. 1931. *Heredity as an Explanation of the Declining Birth Rate*. Cold Spring Harbor, NY: Eugenics Research Association.

10. Draper, W. P. 1960. Memorandum on the New York Plan. *Pioneer Fund Archives*. 18 August.

11. Draper, W. P. 1960. New York Plan. *Pioneer Fund Archives*. 18 August. 11.

12. Kennedy, J. F. 1964. *A Nation of Immigrants*. New York, NY: Harper and Row.

13. Shuey, A. 1958. *The Testing of Negro Intelligence*. Lynchburg, VA: J. P. Bell.

Harry H. Laughlin, M.D. (hon.), D.Sc.
Carnegie Institution
Photo courtesy of the Laughlin Collection,
Northeast Missouri State University Archives

Chapter 2

Harry H. Laughlin

Harry H. Laughlin (1880–1943) was one of the original five directors of the Pioneer Fund and its first president. Laughlin was one of the leading American eugenicists of the first half of the 20th century. He was an advocate of programs involving sterilization of the mentally retarded and habitual criminals, and favored a restrictive immigration policy.

Harry Hamilton Laughlin was born in Oskaloosa, Iowa in 1880, the son of George Hamilton Laughlin, a minister and academic, who held a succession of jobs in minor colleges. Harry Laughlin was brought up in Kirksville, Missouri, where he attended the Kirksville Normal School. He left the school in 1896 and spent the next 10 years teaching school in Iowa. During this time he developed an interest in agriculture and in plant and animal breeding, took a degree in these subjects at the State Normal School in Missouri and graduated Bachelor of Science in 1900. In 1907 he took up a position as lecturer in Agriculture, Botany, and Nature Study at the Kirksville Normal School.

At this time Laughlin was carrying out some experiments in breeding poultry, and in 1908 he wrote to Charles Davenport, the director of the Cold Spring Harbor Research Station on Long Island, for advice on this work. Davenport replied and invited Laughlin to attend a summer course on genetics and eugenics at the Research Station. Laughlin accepted the invitation, and was enthused by the experience. On his return to Missouri he continued his poultry research and began publishing papers on genetics. In 1910 he was invited by Davenport to become superintendent of the newly created Eugenics Record Office at Cold Spring Harbor. In 1917 he obtained a D.Sc. from Princeton for his work in genetics.

THE STERILIZATION ISSUE

Laughlin's first interest in practical eugenics was the promotion of the sterilization of the mentally retarded, the mentally ill, and criminals. In the early decades of the century a number of eugenicists campaigned for the sterilization of these groups on the grounds that they were incapable of rearing children properly and that any children they might have would be likely to inherit the disorders. In 1907 the state of Indiana passed a law authorizing sterilization of some members of these groups. In 1910 the American Breeders Association instructed its Eugenics Section to consider the issue.

Laughlin acted as secretary to the committee set up for this purpose and played a major role in the report, which appeared in 1914. The report estimated that approximately 10 percent of the population had genetically based pathologies of mental retardation, mental illness, and criminal tendency, and recommended the sterilization of these as a means of reducing these social pathologies in future generations. Laughlin wrote a separate report which analyzed sterilization laws which had been passed in 16 American states by 1914 and the legal actions and court decisions which had arisen. The report also contained a draft of a model sterilization law. It recommended that each state should appoint a eugenicist with the responsibility of identifying those for whom sterilization was appropriate and serving the necessary court order. Eight years later Laughlin produced a book *Eugenical Sterilization in the United States*[1] which gave updated information.

THE CASE OF BUCK V. BELL

In the 1920s Laughlin served as an expert witness in the legal case of *Buck v. Bell*. The case arose from the sterilization of Carrie Buck, a mentally retarded 17 year old in Virginia who had been committed to the Virginia Colony for Epileptics and the Feebleminded. She had a mental age of nine years and an IQ of 56. Carrie's mother, Emma, had also been committed to the Colony and was also certified feeble-minded with a mental age of slightly under eight years and an IQ of approximately 50. Shortly before her commitment,

Carrie had had an illegitimate daughter named Vivian, who was taken away from her and placed with foster parents because Carrie was not considered a competent mother.

In September 1924 the director of the facility to which Carrie Buck had been committed decided that she should be sterilized in accordance with the provisions of the State of Virginia Sterilization Statute, which had become law earlier that year. This decision was challenged on behalf of Carrie Buck. The case went first to the Circuit Court in Amherst County, where the sterilization order was upheld. It was then taken to the Virginia Supreme Court of Appeals in 1925, where the order was again upheld, and finally to the United States Supreme Court in 1927, where it was upheld yet again by a vote of 8 to 1. The sterilization was performed, and *Buck v. Bell* has never been overturned.

In these court proceedings Harry Laughlin testified as an expert witness that both Carrie Buck and her mother were feeble-minded, that their feeble-mindedness was hereditary and likely to be transmitted to any further children that Carrie might have. Additional evidence concerning Carrie's daughter Vivian was submitted by Caroline Wilhelm, a social worker, to the effect that Vivian was not a normal baby, and by Arthur Easterbrook of the Eugenics Records Office that a mental test for infants showed that she too was retarded. It was on the basis of the evidence that Carrie Buck herself, her mother, and her daughter were mentally retarded that Justice Oliver Wendell Holmes wrote for the majority of the Supreme Court that sterilization was in the public welfare interest "in order to prevent our being swamped by incompetence," concluding with the memorable phrase, "three generations of imbeciles are enough."[2] *Buck v. Bell* established the legal precedent for the sterilization of the mentally retarded in Virginia. Approximately 7,500 sterilizations were carried out in Virginia through 1972.[3]

Some accounts question whether Carrie Buck was mentally retarded, arguing that the intelligence test was an early one that was unreliable and that in late middle-age she had been seen conversing and assisting in solving a crossword puzzle. These criticisms do not stand up. The intelligence test used was the Terman, the leading test of the period. With a mental age of nine, Carrie Buck would have been able to converse as fluently as the average nine year old child, including making suggestions for the answers to a simple puzzle. There is no reliable evidence to doubt that Carrie Buck had an IQ of approximately 56, well below the threshold of the upper limit for mental retardation of 70.

IMMIGRATION

Laughlin was also involved in the immigration controversies of the 1920s. Until the end of the 19th century, the great majority of immigrants into the United States had come from northwest Europe, principally from England, Scotland, Ireland, Germany, Holland, France, and Scandinavia. The only other sizable group were blacks. From around 1890 onwards the national and ethnic origin of immigrants began to change. Large numbers were coming in from eastern and southern Europe, principally from Russia, Poland, Austria-Hungary, Italy, the Balkans, and Turkey. A number of these were Jews fleeing from the pogroms in Russia and Poland, while others were political dissidents and economic migrants. From 1900 to 1914 immigrants from eastern and southern Europe and from Asia numbered around 800,000 annually.[4]

At this time, many Americans began to feel concerned about the large numbers of these new immigrants. Some of them believed they would cause cultural problems because of the difficulties of assimilating so many peoples of different ethnic backgrounds into a common culture. Some questioned the average intelligence of these new immigrants as well.

Laughlin took this issue up in 1917 in an article in *Eugenical News*, the journal of the Eugenics Research Association which he edited.[5] He proposed new immigration legislation for psychological testing of prospective immigrants for intelligence and temperament in order to screen out those with low IQs and questionable character.

In 1920 the United States Congress was considering measures to restrict immigration, and the House Committee on Immigration and Naturalization appointed Laughlin as Expert Eugenics Agent. He also served on the Eugenics Committee of Congress, chaired by Representative Albert Johnson. On 8 March 1924, Laughlin delivered a lecture to the committee summarizing the evidence about the intelligence of immigrants and estimating that there were approximately 2 million foreign born whites in the United States in intelligence class D whose IQs were below 70.[6]

The congressional hearings on immigration led to the passing of the Johnson-Lodge Immigration Act in 1924. This stipulated that the number of immigrants allowed from any country in any one year was to be limited to 2 percent of the number of American citizens of that national origin recorded in the 1890 census. The effect of this was that there were large immigration quotas from the countries of northwest Europe and small quotas for those of southeast Europe and the rest of the world. In addition, the immigration of Chinese and Japanese was halted.

There has been dispute about how influential intelligence test data were in the passing of the 1924 Act. Mark Snyderman and Richard Herrnstein examined over 600 pages of recorded debate in Congress and found that the intelligence issue was only brought up once. They concluded that the intelligence issue was of little or no significance in passing national origin immigration quotas.[7]

Laughlin's health begin to decline after 1940. He retired from the Eugenics Records Office in that year, and the office closed shortly afterward. Laughlin resigned from the Pioneer board in 1941 and returned to his hometown in Missouri where he died in 1943.

NOTES

1. Laughlin, H. H. 1922. *Eugenical Sterilization in the United States*. Chicago, IL: Psychopathic Laboratory of the Municipal Court.

2. United States Supreme Court 1929. *Records: Buck v. Bell*.

3. Gould, S. J. 1981. *The Mismeasure of Man*. New York, NY: Norton.

4. Lutton, W. and Tanton, J. 1994. *The Immigration Invasion*. Petoskey, MI: Social Contract Press.

5. Laughlin, H. H. 1917. The new immigration law. *Eugenical News*. 2: 22.

6. U.S. House of Representatives, 1924.

7. Snyderman, M. & Herrnstein, R. J. 1983. Intelligence tests and the Immigration Act of 1924. *American Psychologist*. 38: 986-95.

Frederick Henry Osborn, LL.D., D.Sc., Litt.D.
Photo courtesy of the American Philosophical Society

Chapter 3

Frederick H. Osborn

F rederick Osborn (1890–1981) was one of the original five directors of the Pioneer Fund. He served as president from 1941 until 1958. Osborn was one of the leading writers on eugenics of the middle decades of the 20th century.

Frederick Henry Osborn was born in 1890 into a wealthy New York business and banking family. His father, William Church Osborn, was a leading New York corporate lawyer and his uncle, Henry Fairfield Osborn, a paleontologist, was president of the American Museum of Natural History and a supporter of eugenics. Frederick Osborn was educated at Princeton, where he took a course in geology which aroused his interest in human evolution and eugenics. On graduation he went to England to do postgraduate work at Trinity College, Cambridge, and while in England he met several of the leading British eugenicists, biologists, and geneticists.

During World War I, Osborn served with the American Red Cross in France. After the war he worked in New York as an executive in the railroad business. He became vice president of the Detroit, Toledo and Ironton Railroad Corporation and a partner in the banking firm of G. M. P. Murphy and Co.

In 1929 Osborn retired from business to devote himself to eugenics, and he spent the next four years reading widely on eugenics and its core academic disciplines of psychology, demography, and genetics. At the end of this period he edited a set of papers published in 1933 under the title *Heredity and Environment*,[1] and the next year collaborated with Frank Lorrimer, a sociology professor at the American University, in writing *Dynamics of Population*.[2] In 1940 he produced his *Preface to Eugenics*,[3] one of the major treatments of eugenics issues of the middle decades of the century.

In World War II Osborn joined the armed forces and was appointed chairman of the Advisory Committee on Selective Service with the rank of major general. After the war he served as deputy U.S. representative on the United Nations Atomic Energy Commission from 1947 to 1950. He was president of the American Eugenics Society between 1951 and 1955. After the end of World War II Osborn resumed his interests in eugenics. He revised his *Preface to Eugenics* and produced a second edition in 1951. In 1968 he wrote his last book, *The Future of Human Heredity*.[4]

PREFACE TO EUGENICS

Osborn's most important work on eugenics was his 1940 book *Preface to Eugenics* which appeared first in 1940 and in a revised edition in 1951. The principal object of the book was to promote what Osborn called "the eugenic hypothesis." This was that the high fertility of the less intelligent and those with undesirable social personality characteristics, which had been present in America and Europe from the closing decades of the 19th century and had persisted into the middle of the 20th, together with the low fertility of the well educated and the professional classes, would be only a temporary phenomenon which would cease in the near future. This phenomenon was known as dysgenic fertility.

To Osborn it appeared that during the 20th century, knowledge and use of contraception had spread from the well educated to the moderately, and even to some of the poorly educated. The result was that family size had generally declined and the dysgenic fertility of the second half of the 19th century and early decades of the 20th century had become less pronounced.

Osborn believed that this diffusion of family planning knowledge would continue into the second half of the 20th century. Fertility would then become eugenic because the highest fertility would be present among "those persons who make the most effective response to their environment." This was the essence of his "eugenic hypothesis."

Osborn therefore believed that we can be optimistic about the future. Nevertheless, the natural evolution of eugenic fertility could be hastened by eugenicists by working for the improvement of the environmental conditions of the poorer classes. He wrote that:

> a voluntary system of eugenics cannot operate under conditions of extreme poverty, ignorance and isolation. The first eugenics requirement is to raise the poorest environments, in order that parents may have the same freedom of choice as to size of family.[5]

When the living standards and education of the poorest classes are improved, Osborn argued, they would use contraception efficiently. By controlling their family size, their fertility would fall to the same level as that of the professional and middle classes, and when this happens overall fertility would cease to be dysgenic. Eugenicists should therefore work for the improvement of the living conditions and education of the poorer classes. They should also work for the extension of the provision of birth control facilities for those who do not use them to hasten the time when all births are planned. There was nothing else that eugenicists should do at the present stage. Osborn was totally opposed to any form of compulsory eugenics.

However, once the new society had evolved in which everyone uses birth control effectively, Osborn believed that it would be possible for eugenicists to become more active. One thing they could do would be to lobby for the extension of state services to lessen the cost of children to parents with large families.[6] Finally, once the new society has arrived, eugenicists will be able to attempt to promote:

> the introduction of eugenic measures of a psychological and cultural sort which will tend to encourage births among parents most responsive to the possibilities of their environment (that phrase again), and to diminish births among those least responsive, thus bringing about a process of eugenic selection through variations in size of family.[7]

SOCIAL CLASS AND RACE DIFFERENCES

It is not clear in Osborn's 1940 book whether he believed there are social class or racial differences in intelligence or other socially desirable traits. At some points he denied this. He wrote that "science has not produced evidence to support the claim that any nation is racially superior"[8] and that "there is as yet little evidence for a social stratification according to genetic capacities."[9] He maintained that there are genes for high intelligence and sound character in all social classes, but that often their potential is not realized in the poorer classes through lack of education and opportunity.

With regard to race differences in intelligence and personality, Osborn was agnostic in his 1940 book. He thought that this was an issue that should be researched, writing that "it is very important that there should be further scientific studies on the genetic capacities of the different races."[10] Interestingly, in the second edition of his *Preface to Eugenics*[11] he deleted this sentence.

EUGENICS AND DEMOCRACY

A subsidiary theme in Osborn's 1940 book concerned the relationship between eugenics and democracy. Osborn understood that there is a potential conflict between the two because a rigorous eugenics program has to control the rights of the mentally retarded, those with low intelligence, certain criminals, and possibly some with genetic disorders (especially those caused by dominant genes in which approximately half of the children inherit the disease) to have children. Osborn doubted whether a program to restrict the reproductive rights of these groups could be maintained in a democracy. He rejected any proposals of this kind and stated that in eugenic thinking the maintenance of democratic principles must be paramount. In a eugenic democracy, he wrote,

> except in the case of extreme defect, no one would be given or would assume any right to decide who should or who should not have children.... Only in a dictatorship would such power be taken from the mass of the parents and put under arbitrary control. A system of arbitrary control would not be eugenics, but would be simply the application of genetic science to the breeding of specific kinds of men and women.[12]

Osborn was completely opposed to any interference with the right of people to have children, except perhaps for the severely mentally retarded. He hoped for a society in which eugenic fertility would evolve naturally and urged eugenicists to work for such a society in which the better stocks would, of their own free choice, have more children.

Osborn's second point about the relation between eugenics and democracy was that democracy could not survive without an application of eugenics. He thought that the continuation of the dysgenic fertility of the last century would produce a population which would lack the qualities required for the maintenance of a democratic society. He wrote that "a eugenic form of society therefore seems essential to the perpetuation of democracy."[13] Osborn apparently thought that the western democracies had a Catch-22 problem because they couldn't introduce eugenics because this would be undemocratic, but if they did not introduce eugenics they would collapse through the degradation of their gene pool. His hope was that dysgenic fertility would only be a temporary phenomenon and would rapidly be replaced by eugenic fertility. If this were so, democracy would survive and flourish.

THE FUTURE OF HUMAN HEREDITY

In 1968 Osborn published a second book on eugenics, *The Future of Human Heredity*.[14] In this he begins with an account of how in the prehistory of human societies, intelligence and character qualities such as altruism, cooperativeness, and motivation must have had survival value, and this led to a genetic improvement of these qualities over hundreds of thousands of years. He cited evidence indicating that this was still the case in economically undeveloped nations in the first half of the 20th century.

Osborn noted next that the natural positive relationship between socioeconomic status and fertility became negative in the second half of the 19th century in the United States, Europe, and Japan. He considered that this was brought about by the differential use of contraception. He proposed that the population could be divided into three broad groups. First, the higher social classes, the most educated and the most intelligent, who used contraception efficiently and had small families. Second, the middle group who used contraception haphazardly and inefficiently and had some unplanned children and medium sized families. Third, a group who never used contraception, had large families and were the least educated, fell into the lowest social class, and had the lowest IQ.

Osborn cited the results of the 1950 and 1960 American censuses showing that the inverse association between educational attainment and fertility was pronounced among married white American women born between 1901-1905, for whom those with less that 8th grade education had an average of 3.42 children while those with 1-3 years of college had an average of 1.70 children. Among later born cohorts, this differential declined until among the 1926-1930 cohort, the least educated had an average of 3.74 children while the 1-3 year college educated had an average of 3.00. Most of this narrowing of dysgenic fertility was apparently due to the college educated increasing their numbers of children.

Osborn attributed these changes in fertility to economic prosperity. The 1901-1905 cohort reached their child bearing years in the economically uncertain times of the 1920s, when the college educated strictly curtailed their fertility. The 1926-1930 cohort reached their child bearing years in the early post World War II period of economic prosperity and increased their fertility.

Osborn reasserted his "eugenic hypothesis" that this narrowing of educational and social class differentials in fertility could be projected forward and that in the relatively near future the relation of educational level and socioeconomic status with family size would turn positive. He wrote that it was likely that:

> this change to favorable birth differentials between large groups classified by education or by income will take place in the near future in the United States.[15]

In his 1968 book he was more specific on possible eugenic policies than he had been in his *Preface to Eugenics*. He made four recommendations for eugenic policies. These were: (1) an endorsement of Hermann Muller's plan for an elite sperm bank for the use of women whose husbands were infertile; (2) the establishment of more hereditary clinics to give genetic counseling to couples who might be carriers of harmful genes; (3) the promotion of greater awareness among highly intelligent people that they should marry someone of equal ability to increase the probability of producing intelligent children; and (4) the establishment of more birth control advice centers to bring the knowledge and use of birth control to those with poor intelligence and little education. He considered the fourth to be the most important, writing that:

> the most urgent eugenic policy at this time is to see that birth control is made equally available to all individuals in every class of society.

He thought this would be "difficult, but certainly not an impossible task."[16]

Osborn concluded his 1968 book with some speculations about the future. He though that within a relatively short period in the United States couples who were superior in intelligence and character might have an average of four children, while those less favorably endowed might have an average of two. He was hopeful that if this were to happen there would be great gains in the genetic quality of the population. But he asked "Will men in the near future maintain a social, economic, and psychological climate that could bring about such a result?" and answered his own question by affirming that "He certainly has it in his power to do so."

Osborn continued to be interested and active in the eugenics movement up to the publication of his last paper in 1974. He died in 1981.

NOTES

1. Osborn, F. H. (ed.) 1933. *Heredity and Environment.* New York, NY: MacMillan.

2. Osborn, F. H. & Lorrimer, F. 1934. *Dynamics of Population.* New York: Macmillan.

3. Osborn, F. H. 1940. *Preface to Eugenics.* New York, NY: Harper and Brothers; revised edition, 1951.

4. Osborn, F. H. 1968. *The Future of Human Heredity.* New York, NY: Weybright & Talley.

5. Osborn, 1940. 197.

6. *Ibid.* 198.

7. *Ibid.* 198.

8. *Ibid.* 295.

9. *Ibid.* 298.

10. *Ibid.* 78.

11. Osborn, F. H. 1951. *Preface to Eugenics.* (2nd ed.). 122.

12. Osborn, F. H. 1940. *Preface to Eugenics.* (1st ed.). 324.

13. *Ibid.* 299.

14. Osborn, F. H. 1968. *The Future of Human Heredity.*

15. *Ibid.* 31.

16. *Ibid.* 98-99.

John M. Harlan

Chapter 4

John M. Harlan

In addition to Draper, Laughlin, and Osborn, the fourth member of the original board of directors of the Pioneer Fund was John Marshall Harlan. During his 17 years as a Pioneer director, Harlan was diligent in his attention to Pioneer matters, regularly attended board meetings, maintained contact with the other directors individually about Pioneer matters, furnished younger lawyers and clerks from his office to handle administrative matters for Pioneer, and granted use of his office address at 31 Nassau Street, New York, NY as Pioneer's mailing address, all without charge.

Both his father and his grandfather were lawyers. His grandfather, also named John Marshall Harlan, served on the U.S. Supreme Court from 1877 to 1911, and his father practiced law in Chicago. It was in this city that John Harlan was born in 1899. He was educated at Princeton and on graduation he was awarded a Rhodes Scholarship to Balliol College at the University of Oxford. On his return to the United States in 1931, he joined the law firm of Root, Clark, Buckner, and Howland. In World War II Harlan served as head of the Operational Analysis Section of the United States Eighth Air Force, which was composed of handpicked civilians in the fields of mathematics, physics, electronics, architecture and law. For his service, Harlan was awarded the U. S. Legion of Merit and the Croix de Guerre from France and Belgium.

After the war Harlan resumed his legal career. In 1954 Harlan was named to the United States Court of Appeals for the Second Circuit and accordingly felt obliged to resign from the Pioneer Fund board. In 1955 President Eisenhower named him to the United States Supreme Court.

Harlan's grandfather, while a Supreme Court justice, was termed "the Great Dissenter." His was the lone dissent against the *Plessy v. Ferguson* decision, which established the "separate but equal" doctrine that was later overturned by *Brown v. Board*. The younger Harlan, who now asked to be called John M. Harlan to distinguish him from his illustrious grandfather, was to be dubbed "the Great Dissenter of the Warren Court." He is best known for his quote, "You can't find a remedy for everything in the Constitution." It is perhaps the most succinct statement of the philosophy of judicial restraint, which has characterized the post-Warren Burger and Rehnquist courts.

As a justice of the Supreme Court John M. Harlan took a generally conservative position in upholding the principle of equality before the law, regardless of race, and he ruled in favor of the NAACP plaintiffs in the *Brown II* decision and other civil rights cases. He was, however, opposed to racial considerations or quotas in judicial processes or the creation of a separate body of "Negro law."[1] For instance, in 1964 the case of *Swein v. Alabama* was appealed from the Alabama Court to the Supreme Court. The case involved a black defendant who had been found guilty by an all white jury of raping a white woman, and the appeal was launched on the grounds that the jury selection was unfair. Harlan opposed the appeal on the grounds that to overrule the Alabama court would be to establish the principle that juries had to represent proportionately the racial composition of their localities.

During Harlan's tenure on the United States Supreme Court, law professor and noted civil libertarian Norman Dorsen wrote that

> Harlan will be remembered not only for his judicial philosophy, but also for his impressive technical proficiency. He has blended wide learning, a clear and orderly style, and a capacity and willingness to work and rework his opinions. Few Justices have so painstakingly or successfully explained their premises or line of argument, and few in the Court's entire history are as safe as he from the charge that judicial opinions are no more than fiats "accompanied by little or no effort to support them in reason."[2]

Justice John M. Harlan died on December 29, 1971.

NOTES

---

1. Yarbrough, T. E. 1992. *John Marshall Harlan - Great Dissenter of the Warren Court*. New York, NY & Oxford, U.K.: Oxford Press. 236.

2. Dorsen, N., 1969. John Marshall Harlan. In Friedman, L. & Israel, F. L., *The Justices of the United States Supreme Court 1789 - 1969: Their Lives and Major Opinions. Volume IV*. New York, NY: Chelsea House. 2819.

John B. Trevor, Jr.

William D. Miller

May Davie
Photo courtesy of
The New York Times/NYT Photo

Karl Schakel

Marion A. Parrott

Harry F. Weyher

Chapter 5

The Other Directors of the Pioneer Fund

The Certificate of Incorporation of the Pioneer Fund stipulates that the Fund should have five directors, of whom one should be elected president. In addition to Wickliffe Draper, their ranks have included a U.S. Supreme Court Justice, an army general, a president of the American Psychological Association, an atomic scientist, as well as a number of distinguished attorneys. This chapter gives brief biographical accounts of all the Pioneer Fund directors from the establishment of the Fund in 1937 up to 1999, except for Wickliffe Draper, Harry Laughlin, Frederick Osborn, John Harlan, and Henry Garrett, whose biographies appear in individual chapters. The dates shown in parentheses after each name give the years they served on the board. The chapter concludes with an overview of the funding policies adopted by the directors of the Fund.

MALCOLM DONALD (1937-1949)

The fifth member of the initial Pioneer board was Malcolm Donald. He was born in 1877, the son of a Boston lawyer, William A. Donald, and his wife Cornelia (Howes). Malcolm Donald entered Harvard in 1895 and graduated B.A. cum laude in 1899, receiving an M.A. degree the next year. At Harvard he was an outstanding scholar and athlete, playing for the university football team for four years. Donald entered Harvard Law School and became editor of the Harvard Law Review before graduating LL.B cum laude in 1902.

For the next few years, Donald practiced law, including starting a partnership with his friend Jerry Smith under the name of Smith & Donald. In 1907 Smith and Donald became partners in the larger firm of Fish, Richardson and Neave, and left this firm in 1916 to become partners in the firm of Herrick, Smith, Donald, Farley & Ketchum. Both Donald and Smith remained with this firm until their respective deaths. Donald's legal practice was interrupted by World War I, when he joined the War Department in Washington as a civilian. Here, in 1918, he controlled a department employing 1,800 people, with a budget of approximately $1 billion. Despite the importance of his wartime responsibilities and his duties with Herrick, Smith, when a police strike hit Boston in 1919 shortly after his return from the War Department, Donald worked for two months as a traffic officer in the motor corps. During World War II he worked in the Pentagon. Donald served as a director of the Pioneer Fund until his death in 1949.

JOHN H. SLATE, JR. (1941-1946)

John H. Slate, Jr. joined the board of the Pioneer Fund in 1941 on the retirement of Harry Laughlin. John Slate took his first degree at Columbia and a postgraduate degree in law at Columbia Law School. He and colleagues set up the legal firm now called Skadden, Arps, Slate, Meagher and Flom. He acted for a time as editor of the *Columbia Law Review*, and he wrote occasional articles for general circulation magazines. He resigned his Pioneer Fund directorship in 1946.

JAMES P. KRANZ, JR. (1948)

James P. Kranz, Jr. was educated at the University of the South at Sewanee, Tennessee, and at Harvard Law School. He worked as a lawyer for the firm of Root, Clark and served on the board of directors of the Pioneer Fund for part of the year 1948.

HENRY RICE GUILD (1948-1974)

Henry Rice Guild was born in 1896 in Boston, Massachusetts, and was the son of Samuel Eliot Guild and Jessie Motley Guild. He went to Harvard, from which he graduated in 1917 and then entered the United States Naval Academy at Annapolis, Maryland. He was commissioned an ensign in the U.S. Navy. Guild resigned in the summer of 1919, entered Harvard Law School, graduated LL.B. in 1922 and joined the Boston law firm of Herrick, Smith. He was president of the Massachusetts Hospital Life Insurance Company and director of a number of companies and institutions. He was very knowledgeable about ornithology and served for a time as director of the Audubon Society. He joined the board of the Pioneer Fund as director in 1948 and retired in 1974 at the age of 78.

CHARLES CODMAN CABOT (1950-1973)

Charles Cabot was a member of that patrician Boston family of whom it has been written that "the Lowells talk only to the Cabots, and the Cabots talk only to God." He was the son of Henry Bromfield Cabot and Anne McMaster (Codman) Cabot.

Charles Cabot was born in 1900 and educated at Harvard, from which he graduated in 1922. He entered Harvard Law School and after qualifying in 1925, worked in a succession of law firms in Boston.

In the early 1940s Cabot was appointed a Justice of the Superior Court of Massachusetts. He acted as chairman of the Massachusetts Crime Commission and the Massachusetts Bay Transportation Authority, and during World War II was Chief of the Secretariat of the United States Strategic Bombing Survey. In 1950 Charles Cabot became a director of the Pioneer Fund, and he held this position until his retirement in 1973.

JOHN MUNRO WOOLSEY, JR., (1954-1959)

John Munro Woolsey, Jr., was born in 1916, the son of John Munro Woolsey and Alice Bradford (Bacon) Woolsey. He went to Yale, where he was a member of Phi Beta Kappa, and from which he graduated in 1938. He then enrolled in Yale Law School, from which he received an LL.B. degree in 1941. He worked initially in the General Counsel's office of the Board of Economic Warfare. From 1942-1945 he served as a lieutenant in the Navy. In 1945-1946 he worked as a lawyer in the United States Chief of Counsel's office at the Nuremberg war crimes trials of prominent German Nazis. Some of the work involved prosecutions for war crimes in Czechoslovakia, for which the Czech government awarded him the Order of the White Lion. Following the end of the Nuremberg trials, John Woolsey joined the Boston law firm of Herrick, Smith. He served as a director of the Pioneer Fund from 1954-1959.

HARRY F. WEYHER, JR. (1958-PRESENT)

Harry Weyher was born in 1921 in North Carolina and took his first degree at the University of North Carolina at Chapel Hill where he was Phi Beta Kappa. He served with the American forces in Europe during World War II. After demobilization he entered Harvard Law School, where he was note editor of the Harvard Law Review and graduated magna cum laude.

On graduation Weyher joined the law firm of Cravath, Swaine and Moore in New York. In 1951 he served as senior assistant counsel for the New York State Crime Commission under John M. Harlan. In 1954 he became one of the founders of the law firm of Olwine, Connelly, Chase, O'Donnell, and Weyher, at which he specialized in corporate tax and acquisitions. The next year he became the personal lawyer of Wickliffe Draper. He retired from the firm in 1991.

Weyher also taught law as a part-time faculty member at New York University during the 1950s. He wrote a number of articles on legal issues and two books, one of which was an account of how lawyers start out in practice,[1] the second of employee stock ownership.[2]

Weyher joined the board of directors of the Pioneer Fund in 1958 on the resignation of Frederick Osborn, was immediately elected president, and has served in this capacity for over forty years up to the present.

JOHN B. TREVOR, JR. (1959-PRESENT)

John Bond Trevor graduated in engineering at Columbia University in 1931. During World War II he worked as a project engineer at the Naval Research Laboratory in charge of the development of ship-borne anti-aircraft control systems, and wrote two technical manuals for the U. S. Navy. He has also written books on genealogy and yacht racing. His civilian career has been in financial management.

JOHN F. WALSH, JR. (1971-1973)

John F. Walsh took his first degree at Harvard and a post graduate degree in law at Yale. He worked as a partner in the law firm of Whitman, Breed, Abbott, and Morgan.

MARION PARROTT (1973-PRESENT)

Marion Arendell Parrott was born in 1918 in Kinston, North Carolina, the son of Dr. William T. Parrott, a physician, and Jeannette Johnson Parrott. Marion Parrott was educated at The Citadel and at the U. N. C. Law School. He was called up for war service in 1940, graduated from the Field Artillery School and was commissioned in the 101st Airborne Division. He was present at the D-Day landing in Normandy, was wounded and captured in northern France and imprisoned in German-occupied Poland. He escaped in 1945 and made it to Russia. He returned to his unit in France and accompanied the Allied forces in their advance into Germany in the closing stages of the war. He was demobilized holding the rank of major.

Marion Parrott completed his law degree in 1947, after which he practiced law in Kinston. He served in the North Carolina House of Representatives in 1949 and 1951. He joined the board of directors of the Pioneer Fund in 1973.

THOMAS F. ELLIS (1973-1977)

Thomas F. Ellis was educated at the University of North Carolina and the University of Virginia Law School, after which he served as an assistant United States attorney and then co-founded the law firm of Maupin, Taylor & Ellis in Raleigh, North Carolina. He served as a lieutenant in the Navy during World War II. He was special counsel to the North Carolina Governor's Advisory Committee on Education and has been actively engaged in working for the Republican Party including service as State chairman for Ronald Reagan's 1976 presidential campaign, national co-chairman of the 1992 presidential campaign of Jack Kemp, and an advisor to the 1996 presidential campaign of Steve Forbes.

EUGENIE MARY LADENBURG ("MAY") DAVIE (1974-1975)

May Davie was educated at Westover School. In 1930 she married Preston Davie, a distant cousin of Wickliffe Draper. She was active in the Republican Party in New York State during the 1930s through the 1960s. She was a member of the Board of Regents of the National Library of Medicine, 1958-60; trustee of the Taft Memorial Foundation, 1955-64; and a trustee of Adelphi University, 1945-1949. She served briefly as a director of the Pioneer Fund for approximately one year in 1974-1975 until her unexpected death.

RANDOLPH L. SPEIGHT (1975-1999)

Randolph L. Speight graduated at the University of North Carolina and was a partner in the investment banking firm of Shearson, Hamill. He died unexpectedly in 1999.

WILLIAM D. MILLER (1983-1993)

William Dawes Miller graduated cum laude in 1942 from the Carnegie Institute of Technology, now the Carnegie Mellon University, with a bachelor of science degree in mechanical engineering. During World War II, he served in the Special Engineer Detachment of the Manhattan District of the U.S. Army service forces assigned to the Oak Ridge, Tennessee electromagnetic isolation plants, which were responsible for the fissionable uranium isotope 235 which was an essential component in the atomic bomb. Following the conclusion of World War II, Miller pursued a long and successful career in the metals and mining industry. He was employed in various executive positions by Continental Copper and Steel and the Anaconda Company. William Miller died in 1993.

KARL SCHAKEL (1993-PRESENT)

Karl Schakel graduated in aeronautical engineering from Purdue University in 1942. After graduation he founded a helicopter development company and later formed his own engineering company specializing in weapons systems and aeronautical engineering. After selling these business interests to a large conglomerate, Schakel entered the ranching business and has owned or operated farming and ranching properties in Colorado, Texas, and other western states, as well as overseas in 12 countries on five continents.

NOTES

---

1. Weyher, H. F. 1987. *Hanging Out a Shingle: An Insider's Guide to Starting Your Own Law Firm.* New York: Dodd Mead.

2. Weyher, H. F. & Knott, H. 1982. *ESOP: The Employee Stock Ownership Plan.* New York, NY: Commerce Clearing House.

Henry E. Garrett, Ph.D., D.Sc.
Columbia University
Photo courtesy of the American Psychological Association

Chapter 6

Henry E. Garrett

Henry E. Garrett (1894–1973) was a psychologist who worked for most of his career at Columbia University and was Draper's principal scientific adviser from the 1920s until Draper's death in 1972. Garrett himself did not receive any grants from the Pioneer Fund but he recommended Audrey Shuey, R. Travis Osborne, and Frank McGurk to Draper and assisted in finding publishers for their work. Garrett's principal work and expertise in psychology were in the areas of intelligence and statistics.

Henry Edward Garrett was born in 1894 and took his first degree at Richmond University, Virginia. In 1922 he took his Ph.D. in psychology at Columbia University where he joined the faculty. In 1941 he became chairman of the psychology department and retained this position until his retirement in 1956. He then moved to the University of Virginia where he remained until 1964. During World War II Garrett was a member of the Adjutant General's committee concerned with the classification and selection of military personnel. In 1946 he was elected president of the American Psychological Association.

In 1926 Garrett published his first book, *Statistics in Psychology and Education*.[1] This was a textbook which was widely adopted in graduate schools and went through seven editions, the last of which appeared in 1966. Subsequent books were: *Great Experiments in Psychology; Psychological Tests, Methods and Results*, written jointly with M. R. Schneck, a textbook on intelligence and personality; *General Psychology; Elementary Statistics and Psychology and Life*.[2] In 1960 Garrett joined the editorial board of the journal *Mankind Quarterly* as honorary associate editor.

THE DIFFERENTIATION HYPOTHESIS

Garrett's major theoretical contribution to psychology was his "differentiation hypothesis" of intelligence, which he formulated in the 1930s and elaborated in the 1940s[3]. This hypothesis states that the abilities of young children are relatively "undifferentiated," that is to say young children tend to perform consistently at the same level on tests of all cognitive and educational abilities including verbal, arithmetical, spatial, mechanical, memory, and so forth. As children grow older these abilities tend to become more differentiated or independent of one another and the correlations between them decline.

BLACK-WHITE DIFFERENCES IN INTELLIGENCE

Garrett wrote extensively on the issue of the differences in intelligence between blacks and whites in the United States. He first commented on this in his 1933 book *Psychological Tests, Methods and Results*. Reviewing the literature up to that time he noted that blacks obtained lower average scores than whites but was open-minded about the factors responsible. He wrote that:

> By way of summary, it may be said that Negroes tested in the United States are generally inferior to the whites in verbal tests of general intelligence. The Negro is most inferior in tests demanding abstract reasoning and language knowledge and usage; he is equal, and sometimes superior, to the white in tests of memory.... Whether the inferiority shown by the Negro upon mental tests is actually a matter of poorer native equipment rather than the result of more meager environmental opportunity, is still an unsettled question.[4]

This was an early recognition, written in the accepted style of the times, that the magnitude of the black-white difference in intelligence varies for different abilities.

Recent scientific work has confirmed Garrett's conclusion that the black-white average difference is greatest on abstract reasoning and spatial abilities and smaller on simple tasks including those involving short term memory. However, it has not confirmed Garrett's conclusion that the black average equals or exceed the white average on tests of memory. It is now clear that the black average is lower than the white on memory, although less so than for tests of reasoning and spatial ability. This has been shown in the United States by Arthur Jensen[5] and in South Africa by Richard Lynn.[6]

Garrett made a further study of the extent to which black and white average IQs differ according to the type of test by examining the data collected on military conscripts in World War I. The tests given to these conscripts were the Army Alpha, a verbal test, and the Army Beta, a non-verbal test, on which the later Wechsler verbal and performance tests were based. Garrett found that the black-white difference was greater on the non-verbal test than on the verbal.[7] These results confounded the testing critics who maintained that intelligence tests are unfair to blacks because they are less familiar with the verbal content of some tests. Garrett was one of the first to show that verbal tests are not culturally-biased against blacks and that blacks tend to do better on these than on non-verbal tests.

Although in his 1930 book Garrett was agnostic regarding the cause of the black-white difference in intelligence, by the end of World War II he had come to the conclusion that genetic factors were largely responsible. In his 1945 letter in *Science*, therefore, he adopted this position. In 1960 he returned to this problem in a critical essay on the work of Otto Klineberg, a former colleague in the psychology department at Columbia. In 1956 Klineberg had contributed a chapter on "Race and Psychology" to a UNESCO (United Nations Educational, Scientific and Cultural Organization) publication entitled *The Race Question in Modern Science*.[8] Klineberg argued that the difference in average IQ between blacks and whites in the United States could be explained environmentally.

Garrett subjected the arguments to critical analysis. Klineberg's first point was that the intelligence test data collected from military conscripts in World War I had shown that blacks in four of the Northern states had higher average IQs than whites in four of the Southern states. The largest difference was between blacks in Ohio and whites in Mississippi, which amounted to an 8 IQ point advantage for the blacks. Garrett maintained that this could be due to the schooling for blacks in the north being better than that for whites in the south, to selective migration of the more intelligent blacks from the south to the north, or simply from sampling error.[9] He argued that the only reasonable comparison between blacks and whites should be between those in the same state because this provided some degree of control for the quality of the environment. He showed that when the black-white differences within states are examined, whites invariably outperformed blacks, and he pointed out that this was also the case in Ontario, where blacks had attended integrated schools with whites since 1890, and yet scored 15-19 IQ points lower than whites.

Klineberg's second argument was that in infancy blacks perform at the same level on developmental tests as whites and that this shows that the difference in intelligence that emerges later must be due to environmental disadvantages.

Garrett objected that the evidence on this point was inconclusive and cited a study by Myrtle McGraw showing that on the Buhler test of infant development white babies aged 2-11 months averaged 13 DQ (Developmental Quotient) points higher than black babies. Subsequent research summarized by J. P. Rushton has shown that black babies tend to be ahead of whites in early infant development.[10] Hans Eysenck argued that this is consistent with a genetic interpretation of the intelligence difference because of:

> a very general law in biology according to which the more prolonged the infancy, the greater in general are the cognitive or intellectual abilities of the species" and that this law holds for race differences in humans.[11]

The third argument advanced by Klineberg was that blacks and whites produce equal proportions of gifted individuals and cited a study of 8,400 black children in Chicago schools, of whom 103 had IQs above 120. Garrett argued that many of these had pronounced white ancestry and furthermore white children with a mean IQ of 100 would have produced approximately six to seven times this number.

Following Garrett's attack on Klineberg, a counterattack on Garrett was mounted by the anthropologist Juan Comas entitled "Scientific Racism Again" in the October 1961 issue of *Current Anthropology*.[12] This was followed by commentaries by 21 scientists. Garrett responded by writing a summary of the exchange in which he maintained that Comas had failed to address the substantive issues.[13]

THE EQUALITARIAN DOGMA

In 1961 Garrett coined the phrase "the equalitarian dogma" for the assertion, increasingly being made at this time by social scientists in the United States and Europe, that blacks and whites are genetically equal in respect of intelligence and that the lower scores obtained on intelligence tests by blacks were due to environmental disadvantages and the discrimination and prejudice of whites.[14]

Garrett maintained that the equalitarian dogma had begun to take hold from around 1930 onwards among academic social and biological scientists, whose careers were put in jeopardy if they questioned it, and among churchmen and in much of the press, radio, and television. He regarded the equalitarian dogma as an article of faith rather than as having any scientific basis, and he considered the question of how this faith had arisen.

Garrett proposed that three principal factors had been responsible. The first of these was the influence of Franz Boas, an immigrant from Germany who was professor of anthropology at Columbia University from 1899 to 1936. Boas was the leading advocate of the equalitarian dogma during this period. He had first asserted this in his 1911 book *The Mind of Primitive Man*. In the 1938 edition of this book he wrote that:

there is nothing at all that could be interpreted as suggesting any material difference in the mental capacity of the bulk of the Negro population as compared with the bulk of the white population.[15]

Garrett rejected this assertion, pointing to the large number of studies that had consistently shown that blacks perform on average considerably worse than whites on intelligence tests. He also maintained that nowhere in sub-Saharan Africa had people ever constructed an alphabet or written language, created a science or a literature and that there had never been any black man of genius comparable to Aristotle, Galileo, Shakespeare, and so on and that there had never been a black civilization.

Garrett suggested that the second factor responsible for the equalitarian dogma was a reaction against Hitler's racial theories of Nordic superiority and the Nazi extermination of the Jews. He proposed that this had sensitized many to any suggestion of racial differences and explained why many American Jewish intellectuals were prominent in the promotion of the equalitarian dogma. Garrett himself explicitly rejected what he described as "Hitler's cruelties and the absurd racial superiority theories of the Nazis,"[16] but he argued that the Nazi errors did not necessarily mean that all group differences were purely the result of environmental, rather than genetic, factors.

Garrett suggested that the third factor responsible for the rise of the equalitarian dogma was the influence of the political left, for whom racial equality was an article of faith. He was undoubtedly right about this, and many of those who continued to promote the equalitarian dogma (such as Leon Kamin and Stephen Jay Gould) have stated that they identified with liberal-left ideology.

Garrett concluded this article with a statement of his position:

> The weight of the evidence favors the proposition that racial differences in mental ability (and perhaps in personality and character) are innate and genetic. The evidence is not all in, and further inquiry is needed.... At best, the equalitarian dogma represents a sincere if misguided effort to help the Negro by ignoring or even suppressing evidence of his mental and social immaturity. At worst, equalitarianism is the scientific hoax of the century.[17]

This book will adopt Garrett's term, "equalitarian," for the dogma or belief that all significant human differences, between individuals or groups, are purely the result of environmental factors and that genetic factors are absent or trivial. It is useful to have an antithesis to the term, "hereditarian," which represents the position of most Pioneer grantees on the nature-nurture issue. Garrett's "equalitarian" is better than the term, "environmentalist," in that the latter is now generally used to refer to those committed to environmental preservation. Many hereditarians, including Draper, were environmentalists, but none equalitarians.

THE RACE-IQ CONTROVERSY HEATS UP

Garrett also discussed the argument frequently advanced by equalitarians that blacks only do poorly on intelligence tests because they live in impoverished environments and that if blacks were provided with equal opportunity with whites they would perform as well as whites. When the geneticist Theodosius Dobzhansky put forward this argument, Garrett responded that blacks were to some extent responsible for the impoverished environment in which many of them lived, and, as he put it, "man's genetic constitution determines the environment."[18] This was an anticipation of what later became known as genotype-environment active co-variation which states that individuals seek out their own environments based upon their genetic background, which has become an accepted concept in contemporary behavior genetics.

Garrett also addressed the equalitarian argument that the low average IQs and educational attainment of blacks could be explained as a result of their poorer schooling rather than by genetically based lower intelligence. To counter this argument he published data for black-white differences among 10-12 year olds obtained from a city in Virginia.

These data were for 4,425 white children and 1,725 blacks aged 10-12 years tested with the Lorge-Thorndike Test in 1963. The results were that the white children had a median IQ of 102 and the black children of 86, a 16 IQ point difference. Garrett argued that the schools were of the same quality throughout the city, and therefore that the 16 IQ point difference between the black and white children could not be attributed to school effects.[19]

In the 1960s much of the discussion of the issue of black-white differences in intelligence centered on the World War I evidence on military conscripts which showed a 17 IQ point average inferiority of blacks. In 1967 Garrett assembled extensive new evidence on the black-white intelligence differences among military draftees in World War II, in the Korean War, and in 1966 for the Vietnam War. He presented the results in terms of the percentages of blacks whose IQs exceeded the white median. This figure was 14 percent in World War I and 12 percent in the three succeeding war samples. Thus the black-white difference had apparently increased over the period of approximately 50 years, or at best remained constant. Garrett argued that educational provisions had increased greatly over the half century and that, on the equalitarian argument that poor education was the factor responsible for the low black IQ, this should have improved by 12 black IQ points relative to the white. He concluded that the equalitarian argument failed on this point. Furthermore, he noted that the superiority of blacks in some of the northern states to whites in some of the southern states found among World War I draftees was no longer present in any of the succeeding data sets. He calculated that 20 percent of southern whites scored in the superior and very superior IQ categories, as compared with only 9 percent of blacks from six New England states. He concluded that improvements in education nationwide had increased rather than diminished the black-white differential and that:

> the persistent and regular gap between Negroes and Whites in mental test performance strongly indicates significant differences in native ability.[20]

Another point Garrett discussed centered in the good sporting abilities of blacks. He cited evidence that in 1966 10 of the 22 top American football players were blacks although blacks constituted only 11 percent of the population and that among professional athletes as a whole, approximately one third were black. If the equalitarians were right, he argued, in their assertion that blacks are handicapped by their poor environments in respect of intelligence and educational attainment, it is impossible to account for their over-representation in top rank sport. The most probably explanation for blacks' strong abilities in sport, he argued, lay in their genetic constitution, and this was another genetic difference between the races.[21] Recently, substantial new evidence for race differences in athletic ability has been summarized by Jon Entine.[22]

Garrett was extensively attacked by the equalitarians for his conclusion that the low average black IQ is genetically based. In 1961 he was criticized by the anthropologist Juan Comas in the journal *Current Anthropology* and by Ashley Montagu in *Perspectives in Biology and Medicine* asserting that Garrett knew next to nothing about psychology and that:

> I'd put any first-year student of psychology up against Professor Garrett in a test of general psychological knowledge and bet heavily on the student coming out ahead.... Garrett has prejudged the evidence; as one who was born and raised in the black belt, *he knows* that the Negro is inferior, and he will distort, sleight-of-hand, and otherwise deform the facts to suit his argument.[23]

These criticisms were answered by Garrett,[24] and he retaliated with a caustic review of Montagu's book *Human Heredity* in which Montagu repeated yet again his claim that there were no genetic differences between blacks and whites in regard to intelligence. In support of this contention, Montagu cited a study showing that black and white babies perform equally well on simple physical tests. Garrett observed that chimpanzee babies do better than either and that performance on these tests has no predictive validity for subsequent intelligence. He concluded that:

> this book cannot be recommended except as an example of how far science can be prostituted to equalitarian ends."[25]

SOCIAL PROBLEMS

Garrett discussed what he saw as the social problems of the multi-racial society including the high rate of crime of blacks and their poor average level of educational achievement. In regard to crime, he responded to an assertion made by Otto Klineberg that there were no race differences in crime by citing 1954 FBI statistics showing that the black-white ratios for crime were 16:1 for rape.[26] In a subsequent paper he noted that crime data for 2,446 American cities for the year 1960 showed that blacks, who comprised about 10 percent of the population, were responsible for over 50 percent of the crimes of murder, robbery, rape, prostitution, and the illegal possession of weapons and that 54 percent of those executed for murder were black.[27]

Garrett was also concerned about the low average level of educational performance of blacks and the adverse effects this would have on white children following full-scale integration of the traditionally segregated schools in the American South. He believed that the presence of large numbers of blacks in white schools would require a leveling of the curriculum which would hold whites back and leave them bored, while blacks would find the curriculum too difficult and become frustrated. He testified to this effect as an expert witness in the case of *Brown v. Board of Education of Topeka* which came before the U.S. Supreme Court in 1954.

In addition to his psychological research, Henry Garrett served briefly as a director of the Pioneer Fund from 1972 until his death at Charlottesville, Virginia, in 1973.

NOTES

1. Garrett, H. E. 1926. *Statistics in Psychology and Education.* New York: Longmans, Green. 1st ed.).
2. Garrett, H. E. 1930. *Great Experiments in Psychology.* New York, NY: Appleton, Century, Crofts; 3rd ed. Oklahoma City, OK: Century Press, 1981; Garrett, H. E. & Schneck, M. R. 1933. *Psychological Tests, Methods and Results.* New York, NY: Appleton, Century Crofts; Garrett, H. E. 1950. *General Psychology.* New York, NY: American Book. Co. (2nd. ed.); Garrett, H. E. 1956. *Elementary Statistics.* New York, NY: Longmans Green; Garrett, H. E. 1970. *Psychology and Life.* New York, NY: Social Science Press.
3. Garrett, H. E., Bryan, A., & Peel, R. 1935. The age factor in mental organization. *Archives of Psychology.* 176: 1-31; Garrett, H. E. 1946. A developmental theory of intelligence. *American Psychologist.* 1: 372-377.
4. Garrett, H. E. & Schneck, M. R. 1933. *Op. cit.* 204.
5. Jensen, A. R. 1998. *The g Factor.* Westport, CT: Praeger.
6. Lynn, R. & Owen, K. 1994. Spearman's hypothesis and test score differences between whites, Indians and blacks in South Africa. *Journal of General Psychology.* 121: 27-36.
7. Garrett, H. E. 1945. "Facts" and "interpretations" regarding race differences. *Science.* 102: 404-406.
8. Klineberg, O. 1956. Race and psychology. In *Race and Science: Scientific Analysis from UNESCO.* New York, NY: Columbia University Press: 423-452.
9. Garrett, H. E. 1960. Klineberg's chapter on race and psychology: a review. *Mankind Quarterly.* 1: 15-22.
10. Rushton, J. P. 2000. *Race, evolution, and behavior: A life history perspective.* (3rd edition). Port Huron, MI: Charles Darwin Research Institute.
11. Eysenck, H. J. 1971. *Race, Intelligence and Education.* London: Temple Smith. 83.
12. Comas, J. 1961. Scientific racism again? *Current Anthropology.* 2: 303-314.
13. Garrett, H. E. 1961. The scientific racism of Juan Comas. *Mankind Quarterly.* 2: 100-106.

14. Garrett, H. E. 1961. The equalitarian dogma. *Perspectives in Biology and Medicine.* 4: 480-484.

15. Boas, F. 1938. *The Mind of Primitive Man.* New York, NY: Appleton, Century, Crofts. 268.

16. Garrett, H. E. 1961. *Op. cit.* 256.

17. *Ibid.* 257.

18. Garrett, H. E. 1966. Heredity and the nature of man. *Mankind Quarterly.* 6: 239-241.

19. Garrett, H. E. 1964. IQ and school achievement of Negro and White children of comparable age and school status. *Mankind Quarterly.* 5: 45-49.

20. Garrett, H. E. 1967. The relative intelligence of Whites and Negroes: the United States Armed Forces tests. *Mankind Quarterly.* 8: 64-79.

21. Garrett, H. E. 1964. *Op. cit.*

22. Entine, J. *Taboo: Why Black Athletes Dominate Sports and Why We're Afraid to Talk About It.* New York, NY: PublicAffairs.

23. Montagu, M. F. A. 1961. Letter. *Perspectives in Biology and Medicine.* 4: 134.

24. Garrett, H. E. 1962. The SPSSI and racial differences. *American Psychologist.* 8: 260-263.

25. Garrett, H. E. 1963. Review of Ashley Montagu's *Human Heredity. Mankind Quarterly.* 4: 53.

26. Garrett, H. E. 1960. Klineberg's chapter on race and psychology: a review. *Mankind Quarterly.* 1: 15-22.

27. Garrett, H. E. 1963. Misuses of overlap in racial comparisons. *Mankind Quarterly.* 3: 254-256.

John C. Flanagan, Ph.D.
University of Pittsburgh
Photo courtesy of the American Institute for Research

Chapter 7

John C. Flanagan

The first project supported by the Pioneer Fund was an investigation carried out by John Flanagan (1906–1995) on the effect of financial incentives for inducing United States Army Air Corps officers and their wives to have children. Air Corps officers were selected as being a group characterized by high intelligence, sound character qualities, good health, and physical fitness, likely to pass these traits on to their children.

John Clemens Flanagan was born in 1906 in Armour, South Dakota, where his father was a Baptist minister. At the age of 18 he entered the University of Washington in Seattle, where he majored in physics and mathematics. He began teaching in high school in Seattle and at the same time took part-time courses at the University of Washington graduate school. It was at this time that he became interested in psychology. This led him to take a summer course in psychology at Yale, and it was here that he met Edward L. Thorndike, who was giving one of the courses. At the end of the 1920s he obtained a fellowship for post-graduate work on mental measurement at Harvard, where he obtained a Ph.D. in 1934.

Flanagan stayed at Harvard to work for Walter F. Dearborn analyzing data collected in local schools over a 12 year period in the Harvard Growth Study; he subsequently became an associate professor at Columbia University, working with Benjamin D. Wood, director of the Cooperative Test Service of the American Council on Education, where he was in charge of the annual achievement test study. He also studied personality and in 1935 published a book, *Factor Analysis in the Study of Personality*.[1]

FERTILITY OF ARMY AIR CORPS OFFICERS

Flanagan planned the project on the fertility of Army Air Corps officers in two stages. The first was to consist of a survey of the marital status, fertility, fertility intentions, and plans of a sample of the officers. The second was to determine whether providing financial incentives to have additional children would prove effective. Flanagan began work on the first stage in 1937 and completed the study in April, 1938. The report was submitted to the directors of the Pioneer Fund and published a year later.[2]

The survey was based on a sample of 427 officers who had been interviewed regarding their marital status, number of children, number of children planned, the number of children they would have if additional income were available, and the amount of additional income that would have to be provided for them to have another child. The age range of the sample was 20-64 years.

Data were reported first for the marital status and number of children of those in the sample aged 40-55, an age group for whom it can be assumed that fertility is virtually complete. The results were that in this sub-sample which numbered 97, 96.9 percent were married and had an average of 1.72 children. Similar data were obtained from official records for all the Air Corps officers in this age group, who numbered 504. Of these 96.1 percent were married and had an average of 1.46 children. These results showed that the sample was closely representative of all Air Corps officers, and their fertility was well below the replacement level of slightly more than two children for each individual and his wife. This supported the concern felt by eugenicists of the period that elite groups in the United States were not having enough children to maintain their numbers or their proportion of the population. These results were quite similar to those which Frederick Osborn and Frank Lorrimer had reported some years earlier for Harvard and Yale alumni who were found to have had an average of 1.55 children.[3]

The socioeconomic status of the fathers in the sample of Air Corps officers and their wives was also ascertained. The results showed that they came disproportionately from the professional class (26 percent of the officers and 31 percent of their wives), although the professional class was only about 5 percent of the general population. Eight percent of the officers and 3 percent of their wives had fathers who were skilled workers, and 6 percent of the officers and 7 percent of their wives had fathers who were semi-skilled or unskilled, as compared with 29 percent of the population.

The study included questions on the number of children considered ideal, the number planned, the reasons for the number planned being quite low, and the extent to which the couples might have more children if financial incentives were offered. The results were that the sample considered the ideal number of children for the average American family to be 3.24. The ideal for their own family was put at 2.53, though the average expected number for the sample was 2.00. This indicated that the expected number of children was significantly lower than the ideal number, and therefore that the expected number might be raised by the provision of financial incentives.

As to what might induce the couples to have more children, the responses were: increases in income (38 percent), adequate insurance in the event of the officer's death (15 percent), better health care for the wife (15 percent), better housing (4 percent), and less frequent changes of station (5 percent). Since the most frequently cited constraining factor for not having more children was cost, it again suggested that the couples might increase their numbers of children if financial assistance were provided.

Direct questions on the reaction to such financial incentives showed that 30 percent would not have more children whatever the financial incentives were. This left 70 percent who might respond positively. Estimates of the expenses of rearing a child up to the age of 22, assuming a college education, produced a figure of $17,500. The conclusion of the first stage of this study was therefore that the Air Corps officers had only a small number of children, averaging 1.72 and that about 30 percent of them might have additional children if financial incentives were offered them.

FINANCIAL INCENTIVES FOR CHILDBEARING

The project was now ready to move to the next stage. This was to consist of offering the Air Corps families financial incentives to have additional children to see if they would respond. This plan was drawn up in 1938, and in early 1939 a letter went out to all Air Corps officers advising them of the terms of the proposal. These were: (1) that it was only available to officers who already had three or more living children (the reason being that those with fewer than three might well have been planning to have another child anyway, and the object of the plan was to provide incentives for the couples to have more children than they would have had otherwise); and (2) the child had to be born in the calendar year 1940. Air Force officer families meeting these conditions were invited to apply for scholarships to support the education of the child. This would be done by an annuity to be purchased by the Pioneer Fund for each qualifying child which would be paid out in eight annual installments of $500 from the child's 12th birthday onwards. The annuity also had a life insurance provision such that in the event of the child's death a lump sum would be paid to the parents. Thus the total sum provided over the eight years was $4,000, which was of course a much greater sum in 1940 than it became after the inflation of the 1970s and 1980s. In 1940 the average pay of Air Force officers aged 31-34 was $4,500, so the total scholarship payable for the additional child amounted to about one year's salary for this age group. It was estimated that about 25 couples might respond to the plan.

The plan only operated for the calendar year 1940, and the officers had to submit their applications by 1 April 1941. The directors of the Fund met after this deadline to consider the results of the project. It was reported that 11 children had been born who met the terms of the plan and whose fathers had submitted claims and that annuities had been purchased for these at a cost to the Fund of $29,843. Flanagan reported that the rather smaller numbers of children being born than had been anticipated was due principally to the uncertain future occasioned by the outbreak of war, and also to the failure of a number of wives to conceive during the specified period.

Flanagan also considered the question of whether the incentives offered by the Pioneer Fund had had any effect in stimulating the birth of the 11 children, who might have been born anyway. To estimate the effect of the incentives, Flanagan obtained information for the number of children born to officers who already had three children over the preceding ten years, and found that the annual average was 4.1. Thus the trend line of 4.1 children jumped to 11 children in 1940, and it was inferred that this increase was attributable to the financial incentives. The conclusion was therefore that the scholarship plan was responsible for the birth of seven children who would otherwise not have been born. The directors decided not to extend the plan for a further year.

Two conclusions can be drawn from the Pioneer Fund's Air Corps scholarship plan. First, the outcome of seven additional children from the target group cannot be said to have made any significant change in the demographic trend. Probably Draper and his co-directors realized this, and it was one of the considerations that led them to decide not to continue the project. Second, however, the study provided an interesting result in showing that financial considerations do affect family planning among professional couples. Fertility rates have been below the replacement level throughout the economically developed world in the last two decades of the 20th century, and this is probably in part due to financial considerations. The Pioneer study suggests financial incentives can do something to alter this demographic trend.

John Flanagan continued to act as an informal consultant to the Pioneer Fund and offered much useful advice over the course of the subsequent half-century. He went on to have a distinguished career in psychology. During World War II he was head of the Aviation Psychology

program of the Army Air Corps. Here he worked
on psychological tests for the selection of air crews
and human factors in the design of equipment. At
the end of the war he took up an appointment as
professor of psychology at the University of
Pittsburgh.

Flanagan expanded the initial work of the
fertility study by later founding the American
Institute for Research, which carried out over 1,000
research projects. The most important was Project
TALENT, an on-going follow-up investigation of
the vocational attitudes, abilities, and career
objectives of more than 400,000 U. S. high school
students.[4] In 1976 he received the Distinguished
Professional Contribution Award from the
American Psychological Association in recognition
of his work on personnel selection, and in 1993 he
received the American Psychological Foundation
Gold Medal. Flanagan died in 1995.

NOTES

1. Flanagan, J. C. 1935. *Factor Analysis and the Study of Personality*. Stanford, CA: Stanford University Press.
2. Flanagan, J. C. 1939. A study of psychological factors related to fertility. *Proceedings of the American Philosophical Society*. 80: 513-523.
3. Osborn, F. H. & Lorrimer, F. 1934. *Dynamics of Population*. New York: Macmillan.
4. American Psychological Foundation. 1997. APF gold medal awards and distinguished teaching of psychology award. *American Psychologist*. 48: 718.

PART II:
The First Research Programs

Audrey M. Shuey, Ph.D.
Randolph-Macon Woman's College
Photo courtesy of the Randolph-Macon Woman's College

Chapter 8

Audrey M. Shuey

Audrey M. Shuey (1910–1977) was the first person to undertake a comprehensive investigation of all the studies that had been carried out on differences in intelligence between blacks and whites in the United States. She analyzed the evidence in detail, considered its interpretation and concluded that genetic factors were largely responsible for the black-white difference.

Audrey M. Shuey was born in Illinois in 1910. She attended the University of Illinois, from which she obtained her B.A. She proceeded to Wellesley College, where she took an M.A., and then to Columbia University, where she worked for her Ph.D. in psychology under the direction of Henry Garrett. After receiving her Ph.D. in 1931, Audrey Shuey received a Laura Spelman Rockefeller Award in child development. In the 1930s she taught psychology at Barnard College and subsequently at the Washington Square College of

New York University. During this time she married and had two daughters. In 1943 she was appointed professor of psychology at Randolph-Macon Women's College in Lynchburg, Virginia. She received an honorary doctorate from Eastern Illinois State University in 1950.

Because of the demands of raising a family as well as university teaching and research, Shuey did not publish a great deal during the early and middle years of her career. She did, however, publish a few papers. One of these was a study of the different scores of black and white students at Washington Square College on the ACE (American Council Examination), in which she reported that black students obtained a mean score of 170 and whites of 215. The standard deviation of the test was approximately 22, so the white students scored approximately 1.5 standard deviations above the black.[1] It was not until she reached her fifties that Audrey Shuey began work on her magnum opus, *The Testing of Negro Intelligence*.[2]

The first edition of this book appeared in 1958 and summarized 240 studies of black-white differences in intelligence. The second edition published in 1966 surveyed a further 140 studies. Shuey was unable to find a commercial or academic publisher willing to publish her book and turned to Draper for support. Draper financed the first edition of her book privately and the second edition through a grant from the Pioneer Fund.

Shuey organized the studies of black IQ into eight categories. These were investigations of: (1) preschool, (2) grade school, (3) high school, and (4) college students; (5) armed forces personnel; (6) delinquents and criminals; (7) the gifted and the retarded; and (8) racial hybrids. The penultimate chapter consists of a review of the evidence on the higher IQs obtained by blacks in the northern states as compared with those in the south, and an evaluation of the selective migration hypothesis to explain this difference. The book concludes with a consideration of the contribution of genetic and environmental factors in explaining the black-white IQ difference.

STUDIES OF PRESCHOOL CHILDREN

Seventeen studies were reviewed of the intelligence of children between the ages of 2 and 6 years and numbering approximately 1,700 blacks and 13,900 whites. The weighted mean IQs obtained by combining the studies were 94 for blacks and 106 for whites, a 12 IQ point differential. For the studies carried out between 1922 and 1944 the means of the two groups were 96.3 and 105.2, a differential of 8.9 IQ points, while the means of the studies carried out between 1945 and 1965 were 90.8 and 107.3, a differential of 16.5 IQ points. Shuey noted that the black-white differential of 12 IQ points for the entire sample was smaller than that found among older age groups. She suggested that this might be due to unrepresentative sampling arising from the selection of most of the children from kindergarten schools. This would explain why both the black and white means were higher than the respective averages. Alternatively she suggested that the black-white differential might be smaller among preschool children than among older children.

STUDIES OF GRADE SCHOOL CHILDREN

Shuey's review summarized 155 studies consisting of a total of approximately 80,000 black school children in the age range 6 to 12 years. Their combined weighted IQ was 84. Analyzed by geographical region, the mean IQ of those in the southern states was 80.5, and of those in the northern states was 87.9. The higher mean obtained by blacks in the northern states confirmed results of the World War I military draft testing. The data were also analyzed for "overlap," that is, the percentage of blacks scoring above the white mean. For a total of 21,477 blacks, the overlap was 12.3 percent.

The analysis of the children by age showed that the younger black children in grades 1-3 obtained a mean IQ of 83.11 and the older children in grades 4-7 a mean of 84.54. Shuey concluded that there was no tendency for the black-white differential to increase over the 6-12 age range.

A number of the studies Shuey reviewed compared black and white children from similar home environments. The conclusion here was that black and white children whose fathers were matched for socioeconomic status still differed by an average of 12.80 IQ points.

Examining the studies in terms of the kinds of ability being measured, Shuey found that black children did relatively well on tests of rote memory or immediate memory. They did poorly, however, in tests of abstract logical thinking and spatial ability. This confirmed the conclusion reached earlier by Garrett[3] which was also later confirmed in greater detail by Jensen.[4]

STUDIES OF HIGH SCHOOL STUDENTS

Shuey's review covered 55 studies with a total of 13,250 black high school students in the age range of 13-18 years. Their mean IQ was 84.14, almost exactly the same as that of the school children in the age range of 6-12 years. Black children in the southern states obtained a mean IQ of 82.42 and those in the northern states 90.77. There was an overlap among the total sample of 9.7 percent.

STUDIES OF COLLEGE STUDENTS

Studies were collated for a total of 64,640 black college students. The most commonly used test was the American Council Psychological Examination for College Freshmen (ACE), taken by 61 percent of the total sample. A mean IQ was not reported, but the overlap figure of 7.2 percent was given for 7,130 individuals. This is lower that the overlap of 9.7 percent for the high school students, indicating a greater differential in IQ among the college students. Shuey attributed this to the fact that the college tests placed a greater demand on reasoning ability.

STUDIES OF ARMED FORCES PERSONNEL

Reviewed next were the studies of the intelligence of military personnel in World War I and World War II. The World War I data were obtained partly from the verbal Army Alpha test and partly from the non-verbal Army Beta. These two tests were broadly similar to the later Wechsler verbal and performance scales, which were based on them. Shuey converted the two tests into a combined scale and calculated that the mean IQ of the black draftees was 83, in relation to a white mean of 100. She noted that for several reasons this 17 IQ point differential was an underestimate of the true black-white difference, principally because a greater proportion of blacks were rejected by the military because of illiteracy, because there were more white officers who were not included in the testing, and because a greater proportion of whites were exempted from service because they possessed strategically valued skills.

The World War II data were presented principally in terms of the proportion of blacks and whites failing the Army General Classification Test. The results showed much greater rejection rates for blacks. Shuey did not calculate the mean IQs of blacks and whites from World War II military data, but she estimated that the differential was greater than in the World War I data. This was later confirmed by John Loehlin, Gardner Lindzey, and J. N. Spuhler, who calculated the black-white difference in the World War II data at 23 IQ points.[5] Finally Shuey presented data from the 1962 military draft for 235,678 whites and 50,474 blacks who were given the Armed Forces Qualification Tests (AFQT). She found that 15.4 percent of whites were rejected for the draft on the basis of poor scores and 56.1 percent of blacks. This shows that black military conscripts at this time were much more highly selected than white.

STUDIES OF DELINQUENTS AND CRIMINALS

Shuey surveyed the intelligence levels of 3,480 black delinquents from 28 studies and concluded that their mean IQ was 74.5. In 15 of these studies comparable results for white delinquents were reported, and their mean IQ was 80.6. Thus black delinquents had IQs about 10 points below the average black IQ, while white delinquents had IQs about 20 IQ points below those of average whites. She also estimated the incidence of delinquency among black and white adolescents and concluded that it was from two to five times greater among blacks than among whites. This confirmed the earlier conclusion of Henry Garrett.[6]

With regard to adult criminals, Shuey reviewed 16 studies. For a total of 1,670 blacks, she calculated that the mean IQ was 81.3, while for a total of 2,407 whites, the mean IQ was 91.8. Shuey's conclusion that delinquents and criminals typically have below average levels of intelligence and that the incidence of criminal behavior is significantly greater among blacks than among whites, has been confirmed by numerous later studies.

The Science of Human Diversity
A History of the Pioneer Fund

STUDIES OF THE GIFTED AND THE RETARDED

Shuey turned next to studies of the incidence of giftedness and mental retardation, defined respectively as having an IQ of above 140 and below 70. She reviewed 30 studies of the incidence of giftedness among blacks and concluded that it occurred in approximately 0.15 percent of the black population. She concluded that the incidence of giftedness among whites was approximately eight times the incidence among blacks.

For the mentally retarded, Shuey summarized 55 studies and concluded that among a total of 33,979 black children, 16.10 percent had IQs below 70, while among 64,834 white children 2.57 percent had IQs below that figure. These are close to the theoretical expectations of the incidence of retardation in the two populations on the assumption that the respective groups' mean IQs are 85 and 100.

Among the severely retarded, however, the incidence of mental retardation was only approximately twice as great among blacks as among whites. Shuey did not comment on the reason for this. The explanation is that among whites nearly all the cases of severe retardation, those with IQs below 50, are caused by single adverse genes or environmental insults such as brain damage at birth. These causes operate equally on blacks, producing the same incidence of severe retardation. However, among blacks appreciable numbers of those with IQs below 50 comprise the lower end of the normal distribution of intelligence, which is inherited polygenically, whereas the incidence of severe retardation arising from this cause among whites is negligible.

The theoretically expected percentages for whites and blacks with IQs below 50, calculated from mean IQs of 100 and 85, respectively, and standard deviations of 15, are approximately 0.1 percent of whites and 1.0 percent of blacks. The effect of this much greater number of blacks is to approximately double the total incidence of severe mental retardation among blacks as compared with whites.

The value of Shuey's work on the incidence of giftedness and mental retardation among blacks and whites was that it showed that the effect of a difference of about 15 IQ points in the means produced an approximately eight-fold difference in the incidence of giftedness and a six-fold difference in the incidence of mental retardation. These differences could be calculated theoretically but Shuey showed that they were actually present in numerous empirical studies.

STUDIES OF RACIAL HYBRIDS

The next question Shuey analyzed was the intelligence of racial hybrids, black-white mixed-race individuals. She found 18 studies in which these had been identified on the basis of their skin color and in which it was possible to compare those of lighter skin color with those of darker. She found that in 16 of the 18 studies, those with lighter skin color obtained higher IQs than those with darker skins. However, she concluded that "these studies make no important contribution to the problem of race differences in intelligence.[7] This was a curious conclusion, because if there is a genetic component to the black-white IQ difference, it follows that the average IQ of mixed-race individuals should be intermediate those of blacks and whites. Studies of mixed-race individuals thereby provide a test of the genetic hypothesis. The fact that in 16 of 18 studies their mean IQ was intermediate, confirms the genetic hypothesis.

THE QUESTION OF SELECTIVE MIGRATION

Shuey's survey confirmed the World War I draft data showing that blacks in the northern states obtained higher average IQs than those in the south. She noted that two theories had been advanced to explain this finding. The first theory proposed that the more intelligent blacks migrated north, so the higher IQs of northern blacks was due to genetics. The environmental theory advanced by Klineberg proposed that the intelligence difference was due to the better educational opportunities then available in the northern states.[8]

Shuey reviewed the 19 studies of this issue and concluded that black migrants to the northern states were more intelligent on average than those who remained in the south by around 4 IQ points. She also concluded that northern-born blacks scored higher than southern-born blacks by around 8 IQ points. This suggests that about half the IQ difference in favor of the northern blacks was attributable to selective migration of the more intelligent blacks northwards, while the other half was attributable to the better education and socioeconomic conditions of blacks in the northern states.

STUDIES THAT CONTROLLED FOR SOCIOECONOMIC STATUS

Shuey found 42 studies which compared the intelligence of black and white children of the same socioeconomic status or used some associated measure such as living in the same neighborhood or attending the same schools. Surprisingly, she found that controlling for the socioeconomic environment made virtually no difference in the black-white difference. For upper class children she calculated the mean IQs of whites at 111.88 and of blacks at 91.63, while for the lower class children, the mean IQs were 94.22 for whites and 88.19 for blacks. Combining the two SES classes within race gave a black-white difference of 16.14 IQ points, virtually the same as that obtained when SES was not controlled. Shuey noted that the lower-SES white children had a higher average IQ than upper-SES black children (IQs of 94.22 and 91.63, respectively)

Shuey argued that the main inference is that SES differences between blacks and whites could not explain the observed IQ differences, contrary to the assertions of prominent social scientists most frequently put forth at that time. Furthermore, the finding that within each racial group the intelligence of the children is positively related to the socioeconomic status of their parents suggests that intelligence determines social status and that the social classes are to some degree segregated genetically on the basis of intelligence, as asserted by R. A. Fisher,[9] and later by Richard Herrnstein,[10] and Arthur Jensen.[11]

EXPLANATIONS FOR THE BLACK-WHITE IQ DIFFERENCE

At the end of her book, Shuey turned finally to a consideration of the possible explanations for the black-white difference. She reviewed the evidence for the hypotheses that being tested by a white psychologist might depress the scores obtained by black children, that blacks might be less motivated to do well on the tests, or that blacks have a lower sense of personal worth than whites which might adversely affect their performance. She found that none of these hypotheses stood up to examination; all were disproven by the data.

Shuey's conclusion as to the black-white IQ difference was that all the strands of evidence taken together: the consistency of the difference in all groups, in all age ranges and in all geographic locations; the stability of the difference over the half century from World War I to the 1960s; the tendency of racial hybrids to score intermediate blacks and whites; the tendency for the racial difference to be greater on tests of abstract reasoning than on tests of cultural knowledge (contrary to the prediction of the Culture Hypothesis); the failure of the data to confirm the competing purely environmental explanations that blacks were adversely affected by being tested by whites, were less motivated or had lower self-esteem; and the evidence that black children scored lower than whites even when they came from families of the same socioeconomic status, argued strongly for a genetic component to the black-white difference in average IQ. Subsequent work by Jensen and Lynn have confirmed her conclusions regarding the facts of the black-white difference, and these have been accepted in standard textbooks such as Brody.[12]

Audrey Shuey died in Lynchburg, Virginia, in 1977. Later editions of her original *The Testing of Negro Intelligence* (which was printed with support from the Pioneer Fund after commercial and academic publishers had refused because of the subject matter) have now become the standard reference work in the area.[13]

NOTES

1. Shuey, A. 1942. A comparison of Negro and white college students by means of the American Council Psychological Examination. *Journal of Psychology.* 14: 35-52.

2. Shuey, A. 1958. *The Testing of Negro Intelligence.* 1st ed. Lynchburg, VA: J. P. Bell; 1966. (2nd ed.) New York, NY: Social Science Press.

3. Garrett, H. E. 1945. Comparison of Negro and white recruits on the Army tests given in 1917-1918. *American Journal of Psychology,* 58: 480-495.

4. Jensen, A. R. 1985. The nature of black-white differences on various psychometric tests: Spearman's hypothesis. *Behavioral Brain Sciences.* 8: 193-219.

5. Loehlin, J. C., Lindzey, G. & Spuhler, J. N. 1975. *Race Differences in Intelligence.* San Francisco, CA: W. H. Freeman.

6. Garrett, H. E. 1945. *Op. cit.*

7. Shuey, A. 1966. *Op. cit.* 466.

8. Klineberg, O. 1935. *Negro Intelligence and Selective Migration.* New York, NY: Columbia Univ. Press.

9. Fisher, R. A. 1929. *The Genetical Theory of Natural Selection.* New York, NY: Dover.

10. Herrnstein, R. J. 1971. *IQ in the Meritocracy.* Boston, MA: Atlantic-Little Brown.

11. Jensen, A. R. 1973. *Educability and Group Differences.* London: Methuen.

12. Brody, N. 1992. *Intelligence.* New York: Academic Press.

13. Osborne, R. T. & McGurk, F. (eds.) 1982. *The Testing of Negro Intelligence Volume 2.* Athens, GA: Foundation for Human Understanding.

Frank C. J. McGurk, Ph.D.
University of Montevallo

Chapter 9

Frank C. J. McGurk

Frank McGurk (1910–1995) was one of the first to challenge the "equalitarian" argument that the differences in average IQ obtained by blacks and whites in the United States was caused solely by environmental factors. This position had achieved a kind of official status in 1950 when the United Nations Educational, Cultural, and Scientific Organization (UNESCO) issued a statement on race stating that there was no evidence for any genetic differences between the races with regard to intelligence.[1] This view had been assiduously promoted by Franz Boas, Ashley Montagu and others and by the 1950s had become orthodoxy throughout most of the academic world.[2]

Frank C. J. McGurk was born in Philadelphia in 1910 and took his bachelor's degree in 1933 at the Wharton School of Finance and Commerce at the University of Pennsylvania. He enrolled in the University graduate school and obtained an M.A. in psychology in 1937. From 1936 to 1938, he worked as an intern, first in the psychiatric department of the Philadelphia General Hospital, and later at the Institute of the Pennsylvania Hospital for Mental and Nervous Diseases. In 1938 McGurk was appointed clinical psychologist at the Children's Memorial Clinic in Richmond, Virginia. Part of his work involved the assessment of the intelligence of delinquents brought before the Juvenile Courts, and he noted that a large percentage of the blacks were mentally retarded. This led him to carry out a study of the intelligence of blacks and whites in Richmond schools. He tested a total of 1,425 whites and 706 blacks in the age range of 4 to 11 years. The results showed a difference of approximately 16 IQ points between the two groups.[3]

McGurk spent the years 1941-45 in the American army. In 1945 he returned to the Wharton School as an instructor in economics and sociology. Two years later he was appointed instructor in the psychology department at Catholic University in Washington, D.C., where he obtained his doctorate in 1951. In 1949 he worked at Lehigh University as an instructor in psychology and in 1954 was appointed staff psychologist at the United States Military Academy at West Point. He left that post in 1956 to take up an appointment as associate professor of psychology at Villanova University, remaining there until 1962 when he was appointed professor of psychology at Alabama College, later to become the University of Montevallo.

TESTING THE CULTURE HYPOTHESIS

Frank McGurk's major work consisted of a systematic investigation of the argument that the black-white difference in intelligence is solely attributable to environmental factors. He called this the "culture hypothesis" because it attributes the difference entirely to culture and denies any genetic contribution. Exponents of the culture hypothesis such as Otto Klineberg and M. F. Ashley Montagu maintained that blacks were handicapped on the intelligence tests for two reasons.[4] The first of these was that many of the tests required cultural knowledge acquired at school, consisting of questions like "Who wrote Hamlet?" and "How far is it to Chicago?" The culture hypothesis asserted that the schools attended by blacks were not as good as those attended by whites and that this explained their poorer performance on the tests. The second proposition of the culture hypothesis was that blacks had lower socioeconomic status than whites and that this depressed their performance.

McGurk resolved to test these two propositions. The method he adopted was to assemble 226 intelligence test questions and get a panel of 78 psychologists and sociologists to classify them into one of three categories: "most cultural," "neutral," or "least cultural."[5] The results showed that there was good agreement among the judges that 103 items were "most cultural" and 81 items "least cultural." These items were then reduced to a subset of the 37 "most cultural" and 37 "least cultural" matched for difficulty, i.e. the percentage of people able to answer them correctly. These 37 pairs of items were then administered to 2,630 white and 213 black high school seniors. From the large number of whites, 213 were selected who matched the blacks in terms of age, socioeconomic status, being in the same school, studying the same curriculum, and having been enrolled in the same school district since the age of six. This produced very closely matched groups, one black, the other white.

When the black-white differences were examined, two points emerged. First, for the test of closely matched groups, the black-white difference was 7.5 IQ points.[6] Thus, even when blacks were carefully matched to whites for socioeconomic status and school environment, they still scored lower than whites, although the deficit was reduced to approximately half that present on unmatched, representative samples. This is the first point on which the culture hypothesis failed.

Second, on the 37 "most cultural" test items, the black-white difference was 5 IQ points, whereas on the 37 "least cultural" items the black-white difference was 10 IQ points. This is, of course, the opposite of what is predicted by the culture hypothesis and is the second point on which it failed.

THE SECULAR TREND OF THE BLACK-WHITE IQ DIFFERENCE

McGurk turned next to a second test of the culture hypothesis.[7] This concerned the point that the relative economic and social conditions of blacks had improved during the first half of the

20th century. The first extensive data on black-white differences in intelligence were collected in World War I and showed a 17 IQ point gap. McGurk argued that this generation of blacks did suffer from socioeconomic handicaps, including generally poorer schools. By the early 1950s, he argued however, the socioeconomic differences between blacks and whites had decreased (even if they had not disappeared completely), and consequently, according to the culture hypothesis, the difference in IQ should have decreased as well. McGurk's analysis of the data showed that this had not occurred, and therefore that the culture hypothesis had failed once again.

In these two papers McGurk also argued that because intelligence is a powerful determinant of educational performance, blacks would be less able, on average, to compete on equal terms with whites. In many racially integrated schools, he argued, blacks would thus be over-represented in remedial classes and be required to repeat grades in classes with younger whites, and this would be damaging to black self-esteem.

THE 1975 STUDY

In the early 1970s McGurk made a further test of the culture hypothesis, assisted by a grant from the Pioneer Fund. He began with a literature search of all the papers that had been written on American blacks between 1951 and 1970 and found 79 that reported data on black-white differences in intelligence. These studies produced a total of 18,670 blacks. Their relative IQ scores were analyzed in terms of "overlap," i.e., the percentage of blacks scoring above the white median. By definition, 50 percent of whites score above the median, so if the average black IQ is lower than the white average, the percentage of blacks scoring above the white median will be less than 50 percent. McGurk found that the median black overlap derived from all the studies was 15 percent. This is approximately equivalent of 16 IQ points, so that if the mean IQ of whites is set at 100, the mean black IQ is 84. McGurk noted that this figure was closely similar to the mean IQs and percentage overlap obtained from studies carried out earlier in the century, including the WWI military draft data published by Yerkes which showed a 17 IQ point black-white differential.[8] McGurk concluded that there had been no reduction in the black-white IQ difference over a period of approximately half a century, from 1917 to 1951-70. He argued that this half century had seen a large improvement in the economic and social conditions of blacks and that the culture hypothesis would predict a parallel improvement in average IQ. The fact that this improvement had not occurred argued against the culture hypothesis.

McGurk then looked at the data on the black-white differences on verbal and non-verbal tests to test the culture hypothesis proposition that blacks are more seriously impaired on verbal tests. He found that 25 of the studies provided data for both verbal and non-verbal intelligence. These studies contained a total of 3,679 blacks, and the percentage overlap was found to be 19 percent for the verbal tests and 15 percent for the non-verbal. Thus, contrary to the cultural hypothesis, the results showed that blacks performed relatively better on the verbal tests, with their greater cultural content, than on the non-verbal tests, in which the cultural content was less. McGurk concluded that once again the culture hypothesis was contradicted by the evidence.

THE 1982 STUDY

A decade later McGurk returned once again to the problem of the black-white difference in intelligence. This study was carried out in collaboration with R. Travis Osborne (Osborne and McGurk, 1982) and consisted of an update of the second edition of Audrey Shuey's *The Testing of Negro Intelligence.*[9] Shuey's book comprised a review of all the studies of black-white differences in intelligence in the United States from the second decade of the century through 1965. Osborne and McGurk's updated book consisted of a review of the studies published from 1966 through 1979 and followed the general format adopted by Shuey. McGurk's contribution was to review the literature on preschool children, school children, and high school students.

For the preschool children 49 studies were found covering approximately 3,700 blacks and 1,000 whites. McGurk calculated the median IQ of the blacks at 89 and of the whites at 109, and the overlap (the percentage of blacks scoring above the white median) at 9 percent. He noted that the black-white difference in these studies was considerably greater than the approximately 13 IQ points Shuey had found in the earlier studies, which again argued against the culture hypothesis.

FINAL RESOLUTION OF THE CULTURE HYPOTHESIS

Over the course of approximately thirty years McGurk had shown that the culture hypothesis as an explanation of the black-white IQ difference was contradicted by a solid weight of empirical evidence. Contrary to the culture hypothesis, the data consistently showed that blacks did not perform worse on cultural tests and often performed better on these than on tests of abstract reasoning. Why this should be was a puzzling problem which was only finally resolved in McGurk's final paper written jointly with Jensen.[10] The resolution was that blacks perform poorest on tests of reasoning and relatively better on tests of acquired information and memory. Cultural tests are largely tests of this second kind, consisting of questions like "Who fought whom in the American War of Independence?" so blacks tend to perform relatively well on them as compared with tests of reasoning ability. This general principle was established by Jensen, of which McGurk's classic 1953 study became a special case.[11]

For school children, 126 studies were reviewed, and the median IQ was calculated at 89 for blacks and 103 for whites, and the overlap was 15 percent. For high school students 17 studies were found. The overall median IQ of blacks was 90 and for whites it was 108, and the overlap was 12 percent. McGurk's conclusion from these studies was that there had been no reduction of the black-white IQ difference in the period 1966-1979 as compared with the period from 1917-1965. McGurk argued that this disconfirmed the culture hypothesis, which proposed that as the socioeconomic conditions of blacks improved, their average IQ would improve relative to whites.

Frank McGurk retired to Florida, where he died at Pompano Beach in 1995. His over thirty years of empirical testing and disconfirming the culture hypothesis represent a major, but unfortunately often unappreciated, contribution to the sciences of psychometrics and differential psychology.

NOTES

1. UNESCO. 1950. *Statement on the Nature of Race and Race Differences*. Paris, France.

2. Boas, F. 1940. *Race, Language, and Culture*. New York, NY: Columbia Press; Montagu, M. F. A. 1942. *Man's Most Dangerous Myth: The Fallacy of Race*. New York, NY: Columbia Press.

3. McGurk, F. C. J. 1943. Comparative test scores of Negro and white school children in Richmond, Virginia. *Journal of Educational Psychology*. 34: 473-481.

4. Klineberg, O. 1935. *Race Differences*. New York, NY: Harper; Montagu, M. F. A. 1942. *Man's Most Dangerous Myth: The Fallacy of Race*. New York, NY: Columbia Univ. Press.

5. McGurk, F. C. J. 1953. Socioeconomic status and culturally-weighted test scores of Negro subjects. *Journal of Applied Psychology*. 37: 276-277.

6. *Ibid*.

7. McGurk, F. C. J. 1953. On white and Negro test performance and socio-economic factors. *Journal of Abnormal and Social Psychology*. 48: 448-450.

8. Yerkes, R. M. 1921. *Psychological Examining in the United States Army*. Washington, D.C.: Memoirs of the National Academy of Sciences.

9. Osborne, R. T. & McGurk, F. C. J. (eds.). 1982. *The Testing of Negro Intelligence Volume 2*. Athens, GA: Foundation for Human Understanding; Shuey, A. 1966. *The Testing of Negro Intelligence*. (2nd ed.). New York, NY: Social Science Press.

10. Jensen, A. R. & McGurk, F. C. J. 1987. Black-white bias in cultural and non-cultural test items. *Personality and Individual Differences*. 8: 295-303.

11. Jensen, A. R. 1985. The nature of the black-white difference on various psychometric tests: Spearman's hypothesis. *Behavioral and Brain Sciences*. 8: 193-219.

Chapter 10

A. James Gregor

A. James Gregor (born 1929) worked on projects supported by the Pioneer Fund in the early 1960s. He wrote a series of papers on the evolutionary basis of ethnocentrism and the genetic basis of race differences. His trenchant analysis of these questions contributed to keeping the evolutionary/genetic viewpoint on these issues alive during a period when equalitarianism would otherwise have been an almost unchallenged dogma.

Gregor was born in 1929 in New York City and was educated at Purdue University, where he obtained a B.A. in history, and at Columbia University, from which he obtained an Ph.D. in philosophy. He was an instructor in social studies at Columbia from 1952–1958. From 1958 to 1961 he was an instructor at Washington College, Maryland, and in 1961 he was appointed assistant professor of philosophy at the University of Hawaii.

He acted for several years as secretary to the International Association for the Advancement of Ethnology and Eugenics, which was supported by the Pioneer Fund. He received a grant in support of his personal research from the Pioneer Fund in 1962. In the same year he became an associate editor of the journal *Mankind Quarterly*, for which he wrote a number of articles in the early 1960s. He was also a frequent book reviewer for the journal, for which his fluency in German, Italian, and Spanish made him an invaluable contributor.

ETHNOCENTRICISM

Gregor was a strong advocate for the significance of the concept of ethnocentrism, a term first proposed by the American sociologist William Sumner in his classic *Folkways*.[1] The essentials of the concept are that evolution has programmed humans to identify with the groups among whom they are reared, to whom they have favorable attitudes, and to feel varying degrees of lack of sympathy, antipathy or hostility toward other groups. This theory was itself derived from the 19th century English sociologist Herbert Spencer's concepts of "in-group amity" (friendship within groups) and "out-group enmity" (hostility to outgroups),[2] Gregor expounded and elaborated the theory in a number of papers and discussed its implications for the difficulty of achieving harmonious race relations in multiracial societies. In 1960 he set the theory out in a paper delivered at the 19th International Congress of Sociology in Mexico, later published in the conference proceedings.[3] He argued that children have an innate learning disposition to develop feelings of identity with the group in which they are raised, that this leads to a preference for the social, political, cultural, ethnic, and racial characteristics of one's own group, and this has the effect of producing genetic homogeneity of local populations and maintaining the genetic diversity of the human species as a whole. Gregor maintained that many species of social animals have the same propensity to mate with those genetically similar to themselves and that this was further evidence for this having a genetic basis in man.

In 1962 he elaborated the theory further in the light of psychological and sociological research findings. He reviewed research showing that racial awareness, identity, that preference for one's own race typically appears in young children at the age of four or five years, and that among whites negative attitudes towards blacks are strong, irrespective of whether or not they have personal contact with them, or live in the North or South of the United States, and are extremely resistant to change.[4] Gregor set out the social implications of his theory in a review of Ruth Glass's book, *Newcomers: The West Indians in Britain*.[5] This book was a sociological account of the experiences of black immigrants from the Caribbean who had come to Britain following the British Nationality Act of 1948, which provided that when Britain's former colonies became independent states their inhabitants would be permitted to come to Britain and acquire citizenship. By 1960 around a quarter of a million blacks from the Caribbean did so and settled in London and other major cities. Glass described the problems of racial discrimination in housing and employment that they had encountered and the widespread "prejudice" of the British towards them.

Gregor argued that this was to be expected because, in accordance with the principle of ethnocentrism, different races instinctively feel antipathy towards one another. He argued that the races have "racially conditioned aesthetic preferences" to find their own race aesthetically more attractive than other races. Consequently the black immigrants would never become fully integrated into white society because "the physical difference which marks them indelibly as a special class acts as a nucleus around which patterns of avoidance and rejection crystallized." He argued that:

> adjusting a permanently distinguishable minority - a racial as distinct from a cultural minority - to a host country is an incredibly complex task, as the historic experience of India, Australia, South and North America testifies.

He noted that research on school children in Britain had shown that young white pre-adolescents expressed curiosity about blacks, but among adolescents this turned to rejection or, as Ruth Glass had phrased it, "the rubbing off" of tolerance.

The antipathy of the British towards the black immigrants was shown by opinion polls, which found that 71 percent of white British were opposed to racial intermarriage and that 61 percent said they would move house if significant numbers of blacks entered their community. Gregor argued that these feelings were instinctive expressions of ethnocentrism and were "preferences rooted in the very dispositions of men."

As a member of the view that ethnocentrism is a part of human nature, Gregor was predictably enthusiastic about Robert Ardrey's book *African Genesis*[6] when it appeared in 1961. Ardrey's book was a restatement of the theory of ethnocentrism. Ardrey asserted, in Gregor's summary,[7] that:

> Man is conceived as a denizen of relatively closed groups, animated by an innate disposition to favor "similars" and suspect "dissimilars" (i.e. preference and prejudice), to identify with a given territory (i.e., nationalism) and to organise himself in a hierarchical social order (i.e., class and/or racial priority.[8]

Gregor noted the contrast between Ardrey's restatement of the ethnocentrism paradigm and the approach of contemporary psychodynamic theories predominant in sociology and psychology which explained human behavior exclusively in terms of early environmental experiences such as broken homes, premature weaning, unresolved Oedipus complexes, or economic deprivation. He welcomed Ardrey's contribution to the theory of ethnocentrism, at the same time warning him that "he can anticipate little profit and no little abuse."

THE GENETIC BASIS OF PREJUDICE

Gregor's view that prejudice is an expression of ethnocentrism and an evolutionarily programmed human characteristic led him into conflict with the then prevailing view of prejudice as a form of mental disorder, particularly prevalent in whites, that required treatment. He took up this issue in a critique of Marie Jahoda's pamphlet *Race Relations and Mental Health*,[9] published by UNESCO as one of a series designed to combat race prejudice. Jahoda was a British social psychologist who applied Freudian theory to the analysis of social problems. She began by asserting that:

> modern biological and psychological studies of the differences between the races do not support the idea that one is superior to another as far as innate potentialities are concerned. [10]

Gregor countered that this was untrue, citing the work of Audrey Shuey[11] who had recently summarized literally hundreds of studies in the United States showing that blacks score lower than whites on intelligence tests, and the belief of a number of geneticists and anthropologists, including Ronald Fisher,[12] that genetic differences between the races probably do exist.

The body of Jahoda's pamphlet consisted of a Freudian psychoanalysis of "prejudiced" individuals who had come to the conclusion that some races may be superior to others. She argued that their "racial prejudice" is a form of mental disorder that can be understood in Freudian terms. Her theory was that white prejudice against blacks arises from over-powerful sexual urges in the white Id which are "projected" onto blacks. Through this psychopathological process, prejudiced whites repress their own powerful sexual drives, "project" them onto blacks, and come to believe that it is blacks who are oversexed and, in addition, lazy, dirty, childish, foul smelling, and carriers of disease. In a further extension of this thesis, Jahoda explained the South African laws prohibiting sexual relations between the races by the theory that white men lusted after black women because of their oversexed Id and needed assistance in controlling these lusts from the state in the form of criminal sanctions against indulging them. Jahoda also tackled the problem of white prejudice against Jews. Her explanation was that whites identify Jews with their own superegos. Whites perceive Jews as having strong superegos because they are typically well motivated, successful, and law abiding. Prejudiced whites reject this over-strong superego and, by extension, the Jews, with which it is associated.

Having to her own satisfaction diagnosed white "prejudice" against blacks and Jews as a form of mental disorder, she considered the problem of how it might be cured. After some discussion she rejected education, exhortation, satire, and voluntary contacts as unlikely to be effective. She concluded that the only effective remedy would be forced residential integration. She noted that whites and blacks were largely segregated by residence and that opinion polls showed that around 80 percent of whites preferred this. Consequently government action would be required to force residential integration on an unwilling population.

Gregor wrote a withering account of this thesis. He described it as an instance of the:

> dramatization so common to the linguistic and speculative gymnastics of psychoanalytic interpretation

and concluded that:

> it is recommended as illustrative of the sort of material which obscures the nature and the character of contemporary social problems.[13]

Three years later Gregor wrote a further critique of the view of prejudice as a form of mental disorder afflicting whites who prefer to avoid mixing with blacks.[14] He considered the use of this concept in the context of white opposition to school desegregation and having their children educated in racially mixed schools. The liberal view of this opposition was that it stemmed from white "prejudice" against their children mixing with blacks. Gregor argued that the research literature showed that educational standards are lower in integrated schools than in all-white schools and that whites were being rational rather than prejudiced when they preferred to have their children educated in segregated schools. The same was true, he argued, for white preferences for living in all-white neighborhoods which are generally safer than black or racially mixed communities. White preferences of this kind are not a form of mental illness, he argued, but a realistic assessment of the nature of the real world. Later Hans Eysenck was to demonstrate in detail in his book, *The Decline and Fall of the Freudian Empire*,[15] the absurdities of Freudian theorizing of the kind exemplified by Jahoda's pamphlet. He showed that most of it consisted of propositions which were unverifiable or, if they were verifiable, had been shown to be wrong. Nevertheless, this was the kind of argument that was being published by UNESCO at this time as part of its crusade for racial equalitarianism.

RACE DIFFERENCES IN INTELLIGENCE

Gregor also took up the issue of racial differences in intelligence and wrote critiques of the work of two of the leading racial equalitarians, M. F. Ashley Montagu and Juan Comas.

Montagu was an anthropologist who wrote a series of books asserting the equalitarian position.[16] Gregor[17] attacked his superficial scholarship. He showed that Montagu's claim that the Australian Aborigines are just as intelligent as whites was supported merely by an anecdote on their good scholastic abilities written by a 19th century missionary and that Montagu omitted to mention the intelligence testing evidence collected by Stanley Porteus[18] which showed that the Aborigines do very poorly on intelligence tests and have average IQs in the range of 50-70. Gregor also critically reviewed Montagu's book *Man in Process*,[19] which maintained once again that all races are genetically equal and that humans are by nature "good, friendly, co-operative, loving and kind," until they are corrupted by society. Gregor ridiculed Montagu's numerous errors and his ignoring of contrary evidence and concluded that:

> much of what he says is at best only partially true. Much more is confused and confusing and a goodly portion is simply in error.[20]

Gregor[21] also critically examined the work of Juan Comas, another prominent racial equalitarian of the time, in a scornful review of Comas's paper *Racial Myths*.[22] He showed that the paper was riddled with errors, that much of it was plagiarized without acknowledgment, and that many of the claims, such as that the Jews are not a biologically distinct population, were clearly wrong. He concluded that

> this essay by Professor Comas affords a signal illustration of how social science essays should not be written.[23]

THE INTELLIGENCE OF THE AUSTRALIAN ABORIGINES

While he was at the University of Hawaii Gregor met Stanley Porteus, professor of psychology at the university, who had developed a paper and pencil maze test as a measure of intelligence. Porteus had tested samples of Australian Aborigines with his test as early as 1915 and again in 1927 and had found that, in relation to white norms, they had a mental age of 10.5, corresponding to an IQ of approximately 66. Gregor repeated the study and tested 50 adult male Aborigines and found that they had an average mental age of 10.4,[24] virtually the same as Porteus had found almost half a century earlier.[25]

PSYCHOLOGICAL EFFECTS OF SCHOOL SEGREGATION

Gregor turned his attention next to a critique of the liberal campaign for racial desegregation of schools. The background of this issue was that schools in the southern states had traditionally been segregated by race. This was challenged in the legal case of *Brown v. Board of Education* which went to the Supreme Court, and in May 1954 the Supreme Court declared that separate schools for blacks and whites were unconstitutional. The court had based this decision to a considerable extent on the expert testimony of Kenneth B. Clark, a psychologist who claimed his research showed that segregated schooling seriously impaired the personality development of black children. Gregor challenged the evidence on which Clark based this claim. He demonstrated that Clark's evidence rested on a study in which 80 percent of black children in segregated schools in a southern state selected brown as their preferred skin color, whereas in an integrated school in the North, 80 percent of black children preferred a white skin. Gregor argued that this could not be interpreted to show that black children in the South were psychologically damaged by segregated schools.

Rather, it suggested that the southern black children had a better self image than the northern, because they preferred their own skin color, whereas the northern blacks were more dissatisfied with their own skin color. The conclusion he drew was that blacks in the segregated South had better mental health than those in the integrated North, contrary to what Clark had testified before the Supreme Court.

A second issue in the schools desegregation debate was whether desegregation would help to overcome the racial prejudice of whites against blacks. Liberals believed that it would have this beneficial effect because whites were only prejudiced against blacks because they had not mixed with them. If blacks and whites could be educated in integrated schools, it was believed, whites would become friends with blacks, would realize what excellent people blacks are, and their racial prejudice would be cured.

Gregor asserted, in accordance with the principle of ethnocentrism, that the feelings of mutual antipathy between different races are biologically programmed and too deep seated to be removed by the panacea of integrated schooling.[26] He argued that the research literature supported this conclusion. He cited studies carried out in the 1940s by P. R. Hofstaetter which concluded that:

> neither better and more widespread education nor a rise in the standard of living affect racial discrimination.[27]

Gregor noted that this conclusion was confirmed by Charles Stember in a book reviewing the effects of education on racial prejudice. Stember concluded that education did not have the ameliorating effects so often ascribed to it, although the "prejudice" of the more educated sometimes took more subtle forms than that of the less educated. "[T]he influence of education" he concluded:

> ...is more superficial than profound...merely raising the educational level will not necessarily reduce prejudice against minorities...the data suggest that the effects are minimal. [28]

Gregor argued that this was only to be expected because the view that racial prejudice is a form of mental disorder caused by deprivations in infancy and easily treatable was misconceived. Racial prejudice was, on the contrary, intrinsic to the nature of human beings and virtually impossible to eradicate.

Gregor abandoned his interests in human genetics and population problems in 1964, when he resigned from the editorial board of *Mankind Quarterly* and from his position as secretary of the International Association for the Advancement of Ethnology and Eugenics. He returned to his original academic interests in philosophy and history and had a successful subsequent career at the University of Kentucky and later at the University of California at Berkeley, where he became an authority on extremist political philosophies, particularly the origins of fascism in Italy.

NOTES

1. Sumner, W. G. 1906. *Folkways*. Boston, MA: Ginn.

2. Spencer, H. 1897. *Principles of Sociology*. New York, NY: Appleton.

3. Gregor, A. J. 1960. Sociology and the anthropological sciences. *Memoire du 19e Congres International de Sociologie.* 11: 83-106.

4. Gregor, A. J. 1962. The dynamics of prejudice. *Mankind Quarterly*. 2: 79-88.

5. Gregor, A. J. 1961. Review of R. Glass's *Newcomers*. *Mankind Quarterly*. 2: 146-148.

6. Ardrey, R. 1961. *African Genesis*. New York, NY: Atheneum.

7. Gregor, A. J. 1962. Review of R. Ardrey's *African Genesis*. *Mankind Quarterly*. 2: 213-214.

8. Gregor, A. J. *Ibid*. 214.

9. Jahoda, M. 1960. Race Relations and Mental Health. In *Race and Science: Scientific Analysis from UNESCO*. New York, NY: Columbia University Press. 453-491.

10. *Ibid*. 9.

11. Shuey, A. 1958. *The Testing of Negro Intelligence*. Lynchburg, VA: J. P. Bell.

12. Fisher, R. A. 1929. *The Genetical Theory of Natural Selection*. New York: Dover.

13. Gregor, A. J. 1961. Race relations and mental health. *Mankind Quarterly*. 1: 248-252.

14. Gregor, A. J. 1964. Review of C. Stember's *Education and Attitude Change. Mankind Quarterly*. 4: 238-239.

15. Eysenck, H. J. 1990. *The Decline and Fall of the Freudian Empire*. Washington, D.C.: Scott Townsend.

16. Montagu, M. F. A. 1942. *Man's Most Dangerous Myth: The Fallacy of Race*. New York, NY: Columbia University Press.

17. Gregor, A. J. 1962. Notes on a scientific controversy. *Mankind Quarterly*. 2: 166-177.

18. Porteus, S. D. 1917. Mental tests with delinquents and Australian aboriginal children. *Psychological Review*. 24 : 32-42.

19. Montagu, M. F. A. 1961. *Man in Process*. New York, NY: Mentor.

20. Gregor, A. J. 1963. Review of Ashley Montagu's *Man in Process. Mankind Quarterly*. 3: 198.

21. Gregor, A. J. 1961. Comas' chapter on racial myths. *Mankind Quarterly*. 2: 30-34.

22. Comas, J. 1961. Racial Myths. In *Race and Science: Scientific Analysis from UNESCO*. New York, NY: Columbia Univ Press. 13-55.

23. Gregor, A. J. *Op. cit.* 34.

24. Porteus, S. & Gregor, A. J. 1963. Studies in intercultural testing. Perceptual and Motor Skills. 16: 705-724.

25. Porteus, S. D. 1917. Mental tests with delinquents and Australian aboriginal children. *Psychological Review*. 24: 32-42.

26. Gregor, A. J. 1964. Review of C. Stember's *Education and Attitude Change*. Mankind Quarterly. 4: 238-239.

27. Hofstaetter, P. R. 1951. A factorial study of cultural patterns in the U.S. *Journal of Psychology*. 32: 100-112.

28. Stember, C. H. 1961. *Education and Attitude Change*. New York, NY: Institute of Human Relations. 180.

Chapter 11

Robert E. Kuttner

Robert Kuttner (1928–1987) was a biochemist whose primary interest was in the genetics and biochemistry of brain function and behavior, on which he published a number of papers. He was also interested in behavior genetics and racial differences, to which he made a number of significant contributions.

Robert E. Kuttner was born in 1928 and took his first degree in biology, specializing in biochemistry, at the City College of New York in 1950. In the 1950s he worked as a research assistant on the physiology and biochemistry of brain function in the department of zoology at the University of Connecticut, from which he obtained a Ph.D. in 1959. In 1961 he was appointed instructor in biology at the Creighton University School of Medicine in Omaha, Nebraska. In 1967 he moved to the University of Chicago to work as research associate in biochemistry in the department of obstetrics and gynecology. In 1970 he moved to Stanford in California to work with William Shockley for approximately a year. In 1972 he was appointed to a professorship at the University of Southern Mississippi, where he worked for a while as a colleague of Roger Pearson and Donald Swan. In the mid-1970s he made another move to work as a microbiologist at the United States Veterans' Hospital in Waukegan, Illinois. Kuttner served as assistant editor of the journal *Mankind Quarterly* from 1962 to 1978.

In the 1960s Kuttner met Donald Swan and in 1967 they published a joint paper on the problem of the persistence of the schizophrenia gene.[1] Schizophrenia is an impairing illness which, among other things, reduces fertility, and this should have eliminated the gene from the population. Kuttner and Swan discussed why this has not occurred. They considered the theory that schizophrenia is caused by a recessive gene which might be advantageous to carriers of the single recessive but debilitating to those inheriting two recessives and expressing the disease. They decided that this theory was implausible and concluded that schizophrenia probably survives through the social support schizophrenics give and receive from the groups of which they are members.

RACE DIFFERENCES IN PREHISTORIC CULTURAL ACHIEVEMENT

From around 1920 to 1960 the debate on race differences in intelligence centered largely on the differences on intelligence test scores between blacks and whites in the United States and the degree to which these were caused by genetic and environmental factors. On the one side people like Henry Garrett, Audrey Shuey, and Frank McGurk argued that genetic factors were largely responsible for these IQ differences, while on the other side people like Ashley Montagu, Juan Comas, and Theodosius Dobzhansky argued that there were no genetic differences between the races and that the lower scores obtained by blacks on intelligence tests were solely caused by their impoverished environment.

Kuttner introduced two new arguments into this debate.[2] The first was the issue of racial differences in prehistoric cultural achievement. His starting point was the work of the anthropologist F. S. Hulse who had discussed the problem of why early civilizations arose in the river valleys of the Euphrates, Indus, Nile, and Yangtze, but had not appeared in other river valleys in other parts of the world. Hulse argued that this was simply fortuitous. Kuttner pointed out that there was a striking contrast between the achievements of the prehistoric Caucasian, Oriental, and American Indian peoples compared with those in sub-Saharan Africa. He cited the ancient Egyptian, Sumerian, Chinese, and Mayan civilizations which had developed agriculture, the domestication of animals, sophisticated tools, and written languages, and noted that none of these developments had occurred in sub-Saharan Africa. He argued that these racial differences in prehistoric cultural achievements might well reflect differences in intelligence which were found on intelligence tests in the 20th century. Some years later this argument was elaborated by the British anthropologist John Baker who drew up a list of 21 characteristics of early civilization which included those instanced by Kuttner and added the inventions of arithmetic, stone or brick built buildings, cities, money, legal systems, sewage disposal, and a number of other criteria.[3] Baker concluded that all 21 components had been present in the early Caucasian civilizations in Sumeria, Egypt, the Greek Isles, and the Indus valley, and in the early Chinese civilization in the valley of the Yangtze river. He concluded that in the early American Indian civilizations, the Maya had achieved about half of the 21 components of civilization, and the Inca and Aztec societies a little fewer. The Negroid peoples and the Australian aborigines, Baker concluded, had achieved none of them. Baker proposed that these differences in cultural achievement were determined by differences in intelligence, but the basic theory had been formulated 12 years earlier by Kuttner.

INTELLIGENCE OF THE NATIVE AMERICAN INDIANS

Kuttner's second contribution to the study of racial differences in intelligence was his work on the intelligence of the Native American Indians.[4] He wrote a thorough review of the literature and concluded that their mean IQs were intermediate those of blacks and whites. However, their average incomes were well below those of blacks. He cited census statistics for 1959-60 showing that median family incomes were $1,900 for American Indians, $3,233 for blacks and $5,835 for whites. He argued that this showed that low incomes in and of themselves did not depress IQs, because American Indians had lower incomes than blacks but higher IQs. He attributed the low incomes of American Indians to their poor motivation for achievement and cited evidence to substantiate this from the Coleman Report. His conclusion was that blacks had reasonably strong work motivation but low intelligence, and this is the explanation for their low incomes.

RACE DIFFERENCES

In 1960 Kuttner wrote a critique of a UNESCO publication on racial conflict.[5] In 1949 the director-general of UNESCO summoned a panel of eight supposed experts in psychology and the social sciences to consider the problem of racial conflict and prejudice. This panel drew up a Statement on Race which announced that humans possess an inborn drive towards brotherhood and mutual co-operation and that the frustration of this drive resulted in diseased individuals and societies. This statement was submitted to approximately 20 social scientists, who approved it, and the statement was published in July 1950. However, a number of biologists protested. This led to the convening of a new conference in which biologists were better represented and the issuing of a revised statement in which the section asserting that humans are instinctively cooperative was omitted. In reviewing this episode, Kuttner noted that there was a contrary view in biology and the social sciences that group and racial competition and conflict is instinctive to humans and that this was the view taken by Charles Darwin, Herbert Spencer and the anthropologist Sir Arthur Keith. He concluded that the original UNESCO declaration was motivated by a political desire to promote social harmony, and did not express the consensus it purported to represent among scientists. He ended by deploring the politicization of sociology in which what were later to become known as "politically correct" theories were presented as if they were scientifically established.

THE SEX RATIO

Kuttner was also interested in social and biological factors affecting the human sex ratio, that is the proportion of male babies to females. In 1962 he summarized the facts on this question as they were known at the time.[6] These were, first, that during times of scarcity or war, the proportion of male babies falls. Kuttner proposed that this was because the male fetus and neonate are more vulnerable to trauma than the female, so more of them die. He also suggested that this served an evolutionary function because females are more valuable than males for maintaining the size of a population. The second fact about the sex ratio was that with a return to peaceful conditions the number of males born increases relative to females. He suggested that the proximate reason for this was that highly fecund couples tend to produce an excess of males and that the restoration of peaceful conditions would raise fecundity because food supplies would be improved, and stress would be reduced. The evolutionary advantage of this would be that it would restore the excess of males after they had been depleted during the previous dearth. As societies settle down, the normal sex ratio is restored.

RACE AND CRIME

Kuttner was interested in the questions of whether crime has a genetic basis and whether there are genetic differences between the races in criminal tendencies. He wrote a review which began by summarizing the seven studies of criminal behavior among identical and same-sexed non-identical twins, carried out in five countries and consisting of a total of 97 twin pairs. The similarity, or concordance as it is technically called, of the identical twins for crime for the combined studies was 68 percent, whereas for non-identicals it was only 28 percent. This difference indicates a strong heritability for criminal tendencies.[7]

As regards the association between race and crime, Kuttner noted the higher crime rates of blacks in the United States as compared with whites. He summarized the explanation for this advanced by Marvin Wolfgang, a sociologist and criminologist at the University of Pennsylvania.[8] Wolfgang argued that blacks commit more visible crimes, such as murder, armed robbery, rape, and so on, while whites commit less visible crimes like tax fraud, misleading advertising, and dubious business practices. Therefore, according to

Wolfgang, there may not be a real racial difference in crime. Furthermore, Wolfgang argued, poverty and social deprivation are associated with crime, and blacks suffer from these more than whites. There is, therefore, according to Wolfgang, no racial difference in crime or, if blacks do commit crime more than whites, this is because of their impoverished living conditions. Kuttner argued that Wolfgang's arguments were typical of the refusal to undertake a serious analysis of the causes of crime, including the possibility that genetic factors might be involved.

Dr. Kuttner died in February of 1987.

NOTES

1. Kuttner, R. E., Lorincz, A. B. & Swan, D. A. 1967. The schizophrenic gene and social evolution. *Psychological Reports.* 20: 407-412.

2. Kuttner, R. E. 1962. Prehistoric technology and human evolution. *Mankind Quarterly.* 3: 71-87.

3. Baker, J. 1974. *Race.* New York, NY: Oxford University Press.

4. Kuttner, R. E. 1968. Use of accentuated environmental inequalities in research on race differences. *Mankind Quarterly.* 8: 147-160.

5. Kuttner, R. E. 1960. The herd instinct in modern sociology. *Mankind Quarterly.* 1: 105-107.

6. Kuttner, R. E . 1962. Biosocial influences on the human sex ratio. *Mankind Quarterly.* 2: 261-263.

7. Kuttner, R. E. 1966. Crime and race. *Mankind Quarterly.* 7: 103-112.

8. Wolfgang, M. E. 1958. *Patterns in Criminal Homicide.* Philadelphia, PA: University of Pennsylvania.

Chapter 12

Donald A. Swan

Donald A. Swan (1928–1980) wrote a number of articles during the 1960s and 1970s on the heritability of intelligence and physical characteristics, including body size and brain size. He also wrote review papers on race differences in intelligence in the United States and their probable genetic basis.

Donald A. Swan was born in 1928 in New York. He took his first degree in economics at Queen's College, New York. In the 1950s and 1960s he worked as an economist in New York. He took an M.A. in mathematics at Hunter College. He also obtained an M.A. in economics at New York University in 1972. In 1973 he was appointed to a professorship in the Department of Sociology and Anthropology at the University of Southern Mississippi where he was a colleague of Roger Pearson.

In the early 1960s Swan developed an interest in the psychology of intelligence, human evolution, and race differences. He published a number of papers on these issues over a period of some 20 years. During much of this period he assisted Robert Kuttner in his work for the International Association for the Advancement of Ethnology and Eugenics. He was supported by the Pioneer Fund for his work on intelligence and for racial differences in blood groups and other physiological characteristics during the 1970s.

THE GENETICS OF PHYSICAL CHARACTERISTICS AND INTELLIGENCE

In 1964 Swan wrote an extensive review of the evidence on the genetic contribution to physical characteristics and intelligence.[1] He summarized the methodologies for tackling this issue consisting of studies of the differences between identical and non-identical twins reared in the same family, the similarities of identical twins brought up apart, and the similarities of adopted children to their biological and adoptive parents. The review concluded that identical twins are considerably more similar than non-identicals reared in the same family, indicating substantial genetic effects; that identical twins reared apart are closely similar in intelligence (the correlation for the twin pairs being between .70 and .80) and that adopted children resemble their biological parents more closely for intelligence than they resemble their adoptive parents. Taking all the evidence together Swan estimated the heritability of intelligence and physical characteristics including stature, height, and head dimensions at between 50 and 75 percent.

THE BLACK-WHITE DIFFERENCE IN IQ.

In 1965, working in collaboration with school administrator H. M. Roland, Swan published some extensive data on the intelligence of black and white children in the city of Wilmington, North Carolina.[2] They reported the results of all seventh, eighth, and ninth grade junior high school pupils who had been given the Otis Mental Ability Test in 1955. The white children, numbering 2,440, obtained a mean IQ of 99.55 (standard deviation = 13.74) and the blacks, 81.24 (standard deviation = 13.45). The intelligence distributions showed that 18.4 percent of black children obtained IQs in the mentally retarded range below 70, as compared with 1.9 percent of whites. At the other end of intelligence, 1.1 percent of blacks obtained an IQ of 110 and above, compared to 21 percent of whites.

The sample was retested in 1959 and showed a slightly reduced difference of 17.36 IQ points as compared with an 18.31 point difference when the samples were first tested. This was attributed to the greater dropout rate among lower-scoring blacks. The paper concluded by arguing for a genetic role in racial difference on the grounds of evidence that equating blacks and whites for socioeconomic background did little to reduce the differential and that the differential had remained the same from 1918 to the 1950s in spite of considerable improvement in the social and economic living conditions of the American blacks.

RACE DIFFERENCES IN BRAIN SIZE AND FUNCTION

Swan was interested in the problem of possible race differences in brain size and function which might relate to the observed differences in IQ. In the early 1960s he wrote a review of the research evidence on this question.[3] Swan was a proficient linguist and was able to read and summarize a number of studies which had been published in German, Italian, and French as well as in English. His review uncovered six respects in which there was evidence of race differences: (1) overall brain size and weight; (2) the degree of convolution of the cerebral cortex; (3) the relative sizes of different parts of the brain, (4) external shape, (5) detailed cytoarchitecture of the gray matter, and (6) electrical activity (EEGs).

In respect of overall brain size, Swan noted first the evidence suggesting that brain size is positively associated with intelligence. This had been shown in several studies. A relationship between brain size and intelligence is also present in comparisons between species in so far as humans have developed much larger brains than the apes from which they evolved, in order to accommodate their greater intelligence. Swan then summarized nine studies of racial differences in brain size, the first of which was published by the Italian anatomist S. Sergi in 1908. Most of these studies compared blacks and whites. Swan concluded that the average brain size of European whites is approximately 100-175cc (about 10%) greater than that of African Negroes and that a similar difference is present for brain weight. He noted that these are average differences, and there is overlap, i.e. some blacks have larger brains than the white average. This conclusion was ignored or derided by equalitarians like S. J. Gould,[4] but was confirmed some quarter of a century later in a series of analyses carried out by J. P. Rushton.[5]

Swan's survey of the literature also indicated that Asians had brain sizes intermediate Caucasians and blacks, while Australian Aborigines had smaller brain sizes than Negroes at (about 1,180 versus 1,240 grams). Subsequent research has shown that the conclusion regarding the small average brain size of Australian Aborigines was correct, but the average brain size of East Asians was underestimated.

Swan turned next to racial differences in the amount of brain convolution, i.e., the extent to which the surface is smooth as contrasted with extensive valley-like indentations. Considered from an evolutionary perspective, the brains of primitive animals have a smooth exterior surface, but in the more advanced mammals, particularly in monkeys and apes, and especially in man, they are convoluted. Since a convoluted brain accommodates more cortical surface, it allows for the development of a higher level of intelligence. Swan concluded that the research evidence pointed to Caucasians having more highly convoluted brains than Negroes.

The research evidence also, in Swan's view, indicated racial differences in the size of different parts of the brain. The brain consists of four lobes designated the frontal, the occipital (at the back), the parietal (top middle), and the temporal (bottom-middle). The anatomical studies suggested that over and above the black-white differences in total brain size, blacks had relatively small frontal and occipital lobes and larger parietal lobes. It was suggested that the difference in frontal lobe size might be that related to race differences in personality and motivation, because it is known that this part of the brain is involved in long term planning, foresight, and the control of impulsiveness. Swan's review of the research evidence also suggested the presence of differences between Caucasians and Negroes in brain shape, the brains of Negroes being longer and narrower than those of European whites.

Swan also looked at the evidence on the cytoarchitecture of the brain. The brain consists of the cerebral cortex, an outer layer of gray matter, linked by neural fibers to other parts of the brain. The cortex is, on average, about 3 mm thick and has several layers and components. Swan's review of the research literature led him to conclude that the cortex was 15 percent thinner in African Negroes as compared with Europeans and that the average black brain had fewer large pyramidal neurons. He cited a study of these differences by anatomist, F. W. Vint, which concluded that the average brain of adult East Africans resembled that of the average 7 to 8 year-old European child in respect of weight and cytoarchitecture.

The final feature Swan examined was the electrical activity of the brain. The electroencephalogram (EEG) records a variety of so-called brain waves and electrophysiological responses to stimulation. Swan summarized a French study which compared the EEGs of 100 Negro soldiers from the Guinea coast with 3,000 Europeans and reported a number of differences indicative of "neuron immaturity" among the blacks. He also reviewed a South African study comparing the EEGs of 66 Bantus and 72 Europeans which reported that the Bantus showed less electrophysiological reaction to flicker stimulation and several other EEG differences.

Although Swan included a number of little-known studies, his review omitted the work of the American 19th century anatomist Samuel Morton who had assembled a collection of 246 skulls. In 1971 Swan returned to the subject of race differences in brain size and added Morton's conclusions.[6] These were that the average cranial capacities of the races, derived from his collection of skulls and measured in cubic inches, were Caucasian (87), Mongoloid (83), Malay (81), American Indian (80), and Negro (78).

Two years later Swan dealt with the issue of the possible effects of racial differences in body size on the brain size differences.[7] He noted that Philip Tobias, a South African anatomist, maintained that the evidence indicating that Negroes had smaller average brain size than Caucasians was invalid because body size differences were not taken into account. Swan argued that this was wrong. For instance, he cited evidence from the American World War I military draft showing that the average height of black and white recruits was identical at 171 cm. Swan's conclusion that American blacks and whites do not differ in height but do differ in brain size was later confirmed in the National Institute of Mental Health study of approximately 17,000 white and 19,000 black children carried out in the 1980s. This showed that at the age of 7 years black children were a little taller than white, but had significantly smaller head circumference at 51.7 and 50.9 cm, respectively.[8] Swan also cited evidence that Australian Aborigines have an average brain weight of about 1,214 grams, as compared to 1,350 to 1,450 for Caucasians, but do not differ in height.

BLOOD GROUPS AND PERSONALITY

In 1975 Swan began an extensive study of the relation between blood groups, intelligence, and personality. The background of this study was an investigation by Raymond Cattell and his colleagues carried out in Italy and the United States in the early 1960s which had found that individuals with blood type A tended to be more "tender-minded" than those with the other principal blood groups.[9] The general interest of this study was its promise of finding genetic links between physiological characteristics and personality.

Swan's original research in this area consisted of a study of 646 white high school students in Southern Mississippi.[10] Blood samples were taken and analyzed for blood group and intelligence and personality measured by Cattell's 16 Personality Factor Test. The only statistically significant result was that those belonging to Type

O blood group scored significantly higher on the nervous tension personality factor. This was a different result from that found previously by Cattell and his colleagues, so the possible relationships between blood types and personality factors remained inconclusive. In the late 1970s he began an international study of racial differences in the frequencies of different blood groups. He was unable to complete this because of his death from complications arising from diabetes in 1980.

NOTES

1. Swan, D. A. 1964. Human genetics and psychological differences. *Mankind Quarterly*. 4: 197-201.

2. Roland, H. M. & Swan, D. A. 1965. Race, psychology and education: Wilmington, North Carolina. *Mankind Quarterly*. 6: 19-36.

3. Swan, D. A. 1962. Juan Comas on "scientific racism" again? A scientific analysis. *Mankind Quarterly*. 2: 231-245.

4. Gould, S. J. 1981. *The Mismeasure of Man*. New York, NY: Norton.

5. Rushton, J. P. 1994. *Race, Evolution, and Behavior*. New Brunswick, NJ: Transaction.

6. Swan, D. A. 1971. The American school of ethnology. *Mankind Quarterly*. 12: 78-98.

7. Swan, D. A. 1973. Brain weight of the Danes. *Mankind Quarterly*. 14: 85-92.

8. Lynn, R. 1990. New evidence on brain size and intelligence. *Personality and Individual Differences*. 11: 795-797.

9. Cattell, R. B., Young, H. B. & Hundleby, J. D. 1964. Blood groups and personality traits. *American Journal of Human Genetics*. 16: 397-402.

10. Swan, D. A., Hawkins, G. & Douglas, B. 1980. The relationship between ABO blood type and factor of personality among South Mississippi Anglo-Saxon children. *Mankind Quarterly*. 20: 205-258.

R. Travis Osborne, Ph.D.
University of Georgia

Chapter 13

R. Travis Osborne

R. Travis Osborne (born 1913) was responsible for some the first work showing that the differences in educational attainment between blacks and whites increase during the years of secondary schooling, that fertility in relation to intelligence was dysgenic in the United States during the late 1960s for both blacks and whites, and that the heritability of intelligence calculated from the comparison of identical and non-identical twins is approximately the same for blacks as it is for whites.

R. Travis Osborne was born in Cocoa, Florida, in 1913 and brought up in West Palm Beach. In 1933 he entered Piedmont College in Georgia. He transferred to the University of Florida in 1935 and graduated in 1936. After teaching for a year in the public school system in Palm Beach County, he entered the University of Georgia graduate school in 1937 and obtained his master's degree in education in the summer of 1938. In

World War II Osborne joined the Navy and was commissioned. He taught air navigation at the Navy Pre-flight School in Athens, Georgia and at the Naval Air Station in Memphis, Tennessee. He retired from the Navy with the rank of commander in the United States Naval Reserve. After discharge from the Navy, Osborne returned to the University of Georgia to continue his graduate work part-time while serving as a director of the Veterans' Guidance Center and assistant professor in the psychology department, where he taught courses in psychometrics. He organized and directed the Student Guidance Center which later became the Counseling and Testing Center. He received his Ph.D. in educational psychology in 1950. He was made full professor in 1960 and, upon retirement in 1980, professor emeritus of psychology. Osborne's research was supported by the Pioneer Fund from 1964 onwards.

BLACK-WHITE DIFFERENCES IN IQ AND EDUCATIONAL ATTAINMENT

In 1954 Osborne began a six-year longitudinal study of the intelligence and educational attainments of black and white children in an unidentified county in the Southeastern United States.[1] He began by administering tests of reading, arithmetic, and general intelligence, using the California Mental Maturity Scale, to 1,467 white and 876 black children in sixth grade, aged approximately 12 years. The results showed that the average black scores were about 1.6 grades below those of whites in reading, one grade below those of whites in arithmetic, and two grades below whites in intelligence. The children were retested when they were in grades 8, 10 and 12, at the approximate ages of 14, 16 and 18 years, although there were inevitably some losses from the sample as a result of drop-outs during the subsequent testings. In the tenth grade the sample was reduced to 539 whites and 273 blacks, among whom the median IQs were 103 and 81, respectively.

The general finding of the study was that the differences in educational attainment and mental maturity between blacks and whites increased between the ages of 12 and 18. For reading, the black-white difference approximately doubled from 1.6 grades among sixth graders to about 3.2 grades among twelfth graders. In arithmetic the difference approximately quadrupled from sixth grade to twelfth grade. The difference in mental maturity rose from about 2 grades to about 3.8 grades.

At this period southern schools in the U.S. were largely segregated by race and Osborne thought it useful to consider the question of whether white teachers were more qualified or experienced than black teachers, and if so whether this could explain the racial differences in attainment. Possibly contrary to expectation, he found that black teachers were better qualified. They had completed a greater number of years of college training, had more master's degrees and five-year teaching certificates, and higher average salaries. These results seemed to argue against differences in teaching quality as an explanation of the differences in educational attainment.

Osborne concluded by discussing the organizational problems of running integrated schools in which there are such marked average differences in intelligence and attainment between blacks and whites. He pointed out that keeping blacks and whites in the same classes produces awkward problems because of the generally poorer black achievement. If those who performed poorly were made to repeat grades, some 80 to 90 percent of blacks would be held back by late adolescence, producing an unsatisfactory mix in the middle grades of younger whites and older blacks performing at about the same level. Another possible arrangement would be segregation of blacks and whites where both groups could be taught appropriately for their abilities. Osborne's final conclusion was that there was no ideal solution to the problem.

DYSGENIC FERTILITY

One of the principal concerns of eugenicists was that more intelligent people have been having fewer children than the less intelligent since at least the closing years of the 19th century. This became known as dysgenic fertility. With the development of intelligence tests in the first decade of the 20th century it became possible to investigate this issue and determine the extent to which dysgenic fertility existed. Two methods were devised for tackling this question. These were the assessment of the intelligence of individuals in relation to their numbers of siblings and the investigation of the intelligence of adults in relation to their numbers of children. Osborne used both methods to make empirical contributions to the study of these issues.

Studies of intelligence in relation to numbers of siblings had been made since the 1920s and had invariably found that the relationship was negative, i.e., intelligent people tended to have fewer siblings. The interpretation of this inverse result was that, because the intelligence of parents and children is positively correlated, it must arise because intelligent people had relatively fewer children. Osborne was possibly the first to study the strength of the inverse relationship between intelligence and numbers of siblings in whites and blacks separately.[2] In 1969 he tested 1,314 school children, of whom 640 were white and 674 were black in a southeastern county, using the California Test of Mental Maturity. Osborne found that there were negative correlations between intelligence and the number of siblings of -.13 among whites and -.10 among blacks. Aggregating the two races produced a higher negative correlation of -.37, the reason for this being that the blacks had larger average numbers of siblings and lower average intelligence. The differences between the numbers of siblings of more and less intelligent children were strikingly evident at the extremes of the distribution. Among those in the 130-149 1Q range, only 10 percent came from families with five or more children. Among those in the 40-49 IQ range, 62 percent came from families with five or more children.

Osborne's second study of the problem of dysgenic fertility consisted of an examination of the relationship between the numbers of children per 1,000 women aged 15-49 in each of 159 counties in Georgia, and the average IQs of the children in these counties. The correlation between the two was -.49, indicating a significant tendency for low IQs to be present in counties where women had large numbers of children.[3] This was the first major study producing direct evidence for pronounced dysgenic fertility in the United States.

HERITABILITY OF IQ IN AMERICAN BLACKS

Osborne carried out important work on the magnitude of the heritability among American blacks. By the 1960s a sufficient number of studies had been carried out on white American twins to show that the genetic contribution to intelligence is substantial. In 1969 these studies were reviewed by Arthur Jensen, who estimated the heritability of IQ among Europeans and white American populations as .81.[4] However, no similar studies with adequate numbers had been made of blacks. This deficiency was made good by Osborne's Georgia Twin Study.[5] He collected data on the IQs of 123 black twins (76 identical, 47 non-identical) and 304 white (171 identical, 133 non-identical). A number of anthropological and blood tests were used to determine whether the twins were identical or non-identical. The twins were given 12 ability tests. When the tests were combined to give a single index of general intelligence, the heritabilities were .62 for the whites and .70 for the blacks. This difference was not, however, statistically significant, showing that the heritability of intelligence is approximately the same in American blacks as it is in whites.

THE TESTING OF NEGRO INTELLIGENCE VOLUME 2

In 1982 Osborne, in collaboration with Frank McGurk, produced a new edition of Audrey Shuey's second edition of *The Testing of Negro Intelligence*.[6] Osborne reviewed the research literature published between the early 1960s and 1980 on the intelligence differences between blacks and whites among high school students taking the Scholastic Aptitude Test and the American College Test; and among college students, adults not in college, and criminals. Osborne concluded that: on the Scholastic Aptitude Test and the American College Test, blacks scored on average at least one standard deviation (equivalent to 15 IQ points) below whites; that overlap lies between 6 and 10 percent; that the College Admissions Tests correlate about 0.55 with grade point averages; that blacks tend to under-perform in colleges as predicted by their aptitude test scores; and that American Indians, Mexican Americans, Asian Americans, and Puerto Ricans all obtain higher average scores on the college aptitude tests than blacks.

For adults not in college, Osborne reviewed 21 studies showing that the average IQs were highest in whites, intermediate in Hispanics and lowest in blacks and that the black-white difference was around 15 IQ points. Among criminals the average IQ of whites lay between 90-95 and of blacks between 75-80, indicating that in both races criminals score 5-10 IQ points below the average of their racial groups. Hispanic criminals typically obtain IQs intermediate blacks and whites.

Osborne also looked at studies of black-white differences in IQ in the four main regions of the United States. He found that the IQs of both races were a little higher in the northeast and in the south, but the differences were approximately the same in all regions. An examination of integrated schools showed that these had no effect on the black-white IQ difference or on differences in scholastic achievement. Finally, Osborne looked at the decline in Scholastic Aptitude Test scores over time and concluded that much of the decline had taken place in the average score because of the increase in the number of blacks taking the test.

HEAD SIZE AND INTELLIGENCE

The relationship between head size and intelligence has been a matter of interest for at least a century. In studies of this issue, head size has typically been estimated from external measurement of the head, either by taking the head circumference or by measuring the length, breadth, and height of the head and calculating the volume. Head size is assumed to be an approximate measure of brain size. A number of studies of this kind have shown low positive correlations of the order of 0.2 between head size and intelligence. Osborne has added to the research literature on this question by calculating the correlations between head size and intelligence for 476 twins, for which he has found a correlation of approximately 0.21, very close to the average of other studies.[7]

In a further study of head size and intelligence, and working in collaboration with J. P. Rushton, Osborne has analyzed his sample of 236 black and white adolescent twin pairs to ascertain differences in head size between males and females, and between blacks and whites, and also to estimate the heritability of head size.[8] The results of the study were that males had larger average head size than females and whites had larger average head size than blacks. These differences were present both absolutely and after adjustments had been made for body size. There were also significant heritabilities for head size.

NOTES

1. Osborne, R. T. 1962. Racial differences in school achievement. *Mankind Monographs*. No. 3.

2. Osborne, R. T. 1970. Population pollution. *Journal of Psychology*. 76: 187-191.

3. Osborne, R. T. 1975. Fertility, IQ and school achievement. *Psychological Reports*. 37: 1067-1073.

4. Jensen, A. R. 1969. How much can we boost IQ and scholastic achievement? *Harvard Educational Review*. 39; 1-123.

5. Osborne, R. T. 1978. Race and sex differences in heritability of mental test performance: a study of Negroid and Caucasoid twins. In Osborne, R. T., Noble, C. & Weyl, N. (eds.) *Human Variation*. New York, NY: Academic Press.

6. Osborne, R. T. & McGurk, F. C. J. 1982. *The Testing of Negro Intelligence, Vol. 2*. Athens, GA: Foundation for Human Understanding.

7. Osborne, R. T. 1992. Cranial capacity and IQ. *Mankind Quarterly*. 32: 275-280.

8. Rushton, J. P. & Osborne, R. T. 1995. Genetic and environmental contributions to cranial capacity in black and white adolescents. *Intelligence*. 20: 1-13.

Ernest van den Haag, Ph.D.
Fordham University

Chapter 14

Ernest van den Haag

Ernest van den Haag (born 1914) is a political philosopher who wrote in the 1950s and 1960s on the issue of school desegregation. Later he contributed to the debates on affirmative action, and he has also written on the problem of race differences in intelligence.

Ernest van den Haag was born in the Netherlands in 1914 in The Hague. Shortly after his birth, his family moved to Germany, where they lived throughout World War I. In the early 1920s the family moved to Italy, where van den Haag went to school and university and studied law. While at the university van den Haag became politically active and opposed to Mussolini's fascist regime. In 1935 he was arrested and imprisoned for two years. On his release from prison in 1937, he went to France. When Germany invaded France in 1940, van den Haag escaped via Spain to Portugal and then to the United States. Shortly after his arrival in the United States, van den Haag gained admission to the State University of Iowa and in 1942 obtained a master's degree in economics and sociology. Between 1942 and 1945 he worked as an analyst at various government agencies. After the end of the war, he was appointed to an instructorship at New York University, from which he obtained his Ph.D. in 1952. In the 1950s he held a Guggenheim fellowship and was a visiting professor at Vassar College; the School of Criminal Justice at the State University of New York, Albany; the University of Minnesota; and the University of Colorado. In 1955 he was appointed adjunct professor of social philosophy at New York University. He also taught social philosophy at the New School for Social Research. In 1978 he was appointed Olin Professor of Jurisprudence and Public Policy at Fordham University, where he remained until his retirement. In addition to his teaching career, van den Haag has been a prolific author and practiced psychoanalysis through the early 1980s.

The fundamental principle in van den Haag's philosophy has been his belief in the primacy of individual freedom over state coercion. This led him to enter the debate on enforced school desegregation and subsequently on affirmative action, both of which he opposed on libertarian and constitutional grounds. The controversies over these issues led him to take up the question of effects of race differences in IQ on school desegregation. Van den Haag received a grant from the Pioneer Fund to work on these issues in 1964.

SCHOOL DESEGREGATION

In many of the U.S. southern states there were separate schools for black and white children until the 1950s. This was challenged in the courts, and in 1954 the Supreme Court ruled in the case of *Brown v. Board of Education* that segregated public schools are unconstitutional and must be desegregated.

Van den Haag was one of the few legal scholars who questioned this decision. In 1956 he produced a book, *Education as an Industry*,[1] in which he argued that the principle of personal freedom and free association required that whites and blacks should be permitted to have segregated schools, if that is what they wish, provided that integrated or what van den Haag called "congregated" schools are available. He pointed out that this practice had been successful with separate schools for boys and girls and predicted that coerced desegregation would not work well. The desire of whites to have their children educated in all-white schools may be based on a prejudice, but van den Haag argued that prejudices are not illegal. He drew a parallel with single-sex colleges (such as Smith College) which many people do not approve of but are nevertheless permitted because some students prefer them. A year later van den Haag produced a second book, *The Fabric of Society*, written jointly with Ralph Ross.[2] In this, they criticized the Supreme Court decision mandating desegregation on libertarian grounds. They argued that voluntary segregation between sexes, religions, or races should be allowed as a basic element in the individual's right to freedom of association, as long as educational institutions existed that were available for each of the groups, where enrollment was voluntary and the facilities were equal in quality, as measured by equal public expenditure per pupil.

In 1960 van den Haag critically examined the psychological testimony given by Kenneth Clark in the lower courts and endorsed by the Supreme Court in *Brown v. Board*.[3] Clark had carried out a study which purported to show that black children in segregated schools suffered psychological damage. Clark testified before the Court and described how he had given black children black and white dolls and asked them which they preferred, which were the nicer dolls and which dolls were most like themselves. The children favored and identified with the white dolls. Clark argued that this indicated personality damage.

Van den Haag's response argued that in American culture, as in many others, white is the color of purity and hope, while black is the color of evil and despair, and also that children are afraid of the dark. Consequently, children would tend to prefer the white dolls and to identify with them. He also pointed to numerous methodological errors in the experiment and still more in its interpretation. What Clark actually found was that preferences for white dolls occurred in 29 percent of black children in segregated schools and in 39 percent of black children in integrated schools. This suggests that black children attending integrated schools alongside whites suffered greater personality damage than those who attended segregated black schools. Clark failed to tell the court this feature of his results. Van den Haag wrote he was therefore

> forced to the conclusion that Professor Clark misled the courts. If Professor Clark's tests do demonstrate damage to Negro children, then they demonstrate that the damage is *less* with segregation and *greater* with congregation. Yet, Professor Clark told the Court that he was proving that "segregation inflicts injuries upon the Negro" by the very tests which, if they prove anything, which is doubtful, prove the opposite.[4]

AFFIRMATIVE ACTION

Van den Haag has been a long standing opponent of affirmative action procedures to increase the numbers of blacks in universities and in employment. In 1965 he wrote that it could not be assumed that the under-representation of blacks is due to discrimination and that therefore what he called "reverse discrimination" is not a legitimate demand.[5] He argued that this was an obvious inequity and that in the long run it would weaken the Negro cause by perpetuating the common stereotype that blacks are invariably less qualified than whites, and by showing up the poor performance of blacks when they are appointed to positions for which they are not qualified. Furthermore, it would undermine the self-esteem of blacks appointed to desirable positions, who would inevitably wonder whether they only secured their jobs because they are black.

Van den Haag argued that whites who advocate affirmative action are probably trying to atone for what they regard as the injustices their ancestors perpetrated against blacks. He argued that this is irrational because current discrimination in favor of blacks does not repair injuries done by whites to blacks in the 19th century, and most injustices are not corrected by making innocent people suffer.

Van den Haag also opposed the extension of affirmative action demanded by a number of black leaders for the kinds of racial integration that can be called coerced congregation or "compulsory togetherness." This consists, for instance, of compelling whites to eat with Negroes, to go to school with them, or to live in their neighborhood.[6] He argued that such demands are "neither legitimate nor fruitful" and that "if gratified, they are more than likely to aggravate the discomfort and strain from which Negroes suffer in American society." He argued that many whites are emotionally unable "to accept Negroes as individuals"[7] and that the courts and the government cannot do anything about this. Human beings, he asserted, have a preference for their own ethnic and national group, and this is "so deeply rooted that mere mixing of groups is unlikely to extinguish it."[8] Even white liberals who oppose segregation and discrimination are often "unable to do the one thing they know Negroes want: accept them as persons and individuals."[9] Many white liberals are hypocrites because they profess to have warm feelings for blacks which they do not really possess. When blacks attack white liberals, these endeavor to assuage their guilt by pressing for "new and quite often unreasonable acts of atonement in favor of Negroes."[10]

In 1976 van den Haag returned to the issue of affirmative action in college admissions and discussed it in terms of its social usefulness and its justice.[11] So far as social usefulness is concerned, he argued that affirmative action tends to exacerbate social divisions by strengthening group identities. Furthermore, the proper function of colleges is to educate the most talented for their future occupations. When the less talented are given preference in admissions, they are helped to enter occupations for which they are less fitted than others who have been excluded under affirmative action admissions procedures. This leads to inefficiency in the allocation of individuals to jobs for which they are most suited, and society as a whole is damaged. Van den Haag concluded that:

> I can find no utilitarian justification for preferential admission of previously underrepresented national, racial or sexual groups. It could be dysfunctional by favoring persons less able to learn and teach than those rejected, thus causing society to be served by the less able in the profession education prepares for.[12]

Van den Haag turned next to the justice argument for affirmative action. He argued that justice requires that colleges should admit the best qualified. Although it was unjust that in the past colleges violated this principle by the preferential admission of whites, males, the sons of alumni and so on, this injustice cannot be rectified by discriminating against these groups. Moreover, the present generation of young whites and males cannot justly be discriminated against in college admissions because some of their fathers and grandfathers benefited from preferential admissions. As a matter of general principle, current generations cannot be made to bear responsibility for discriminations imposed by past generations. Van den Haag asked rhetorically whether today's Italians can be held responsible for the wrongs done by Caesar when he invaded Gaul, or today's Frenchmen for the wrongs done by Napoleon. Similarly, today's American whites cannot be held responsible for the wrongs done by their ancestors to native American Indians. He concludes that:

> Those not responsible for it cannot be asked to compensate for damage they did not cause. Justice requires the cessation of discriminatory practices, including discrimination in admissions, but no reversal.[13]

Van den Haag considered next the assertion frequently made by liberals that the under-representation of blacks in high status universities must be due to discrimination. He argues that this is not so, because the abilities required for college admission are not equally distributed in all races. Only a minority of blacks actually gain from affirmative action in college admissions. However, even for these the gains may be more apparent than real. The blacks admitted in spite of low grades may fail the course and be worse off than before. Or the universities may lower their standards to ensure that the blacks pass, but then these blacks will tend not to perform well in their professions, and this will reinforce white views that blacks are less competent than whites. Or, even if the blacks do succeed, they will be uncertain as to whether this is due to their own talents or to preferential affirmative action. Furthermore, once affirmative action is accepted, there are difficulties in determining how far it should be implemented. If there is affirmative action for blacks, there must be affirmative action also for Hispanics. But what about Portuguese and Brazilian ethnic groups in the United States? Will there have to be affirmative action for these as well? Van den Haag argued that once society goes down the road of affirmative action it becomes enmeshed in innumerable arbitrary classifications and distinctions, and these would require intolerable bureaucratic interference over the freedom of institutions to control their own affairs. Van den Haag ended where he started out: in the primacy of individual freedom over bureaucratic coercion.

RACE DIFFERENCES IN INTELLIGENCE

In 1964 van den Haag turned to the relevance of race differences in intelligence for the issue of school desegregation. He argued in an interview published in the *National Review* that because blacks and whites differ in intelligence, they are

most effectively taught separately.[14] Integrated
schools, he argued,

> would impair the education of Negro and white
> children ... the needs of Negro children would be
> best met i.e., to their advantage and without
> disadvantage to others - by separate education. [15]

His reason for supporting racially segregated
education was that children can be taught most
effectively when they are separated by ability. Asked
whether this would not imply that an exceptionally
intelligent black child should be taught with whites,
van den Haag replied that "this would demoralize
the remaining Negro children and could be hard
also on the transferred child" because the whites
might be hostile to the odd black child. He added
that the instruction given in black schools should
be specifically designed to overcome the
disadvantages of culturally deprived home
environments and that this could not be done
except in separate schools.

Van den Haag went on to discuss the
possibility of genetic differences between blacks and
whites. He defended intelligence testing as a valid
technology for the measurement of genetic
differences and cited the data on the close similarity
of the IQs of identical twins reared in different
families as evidence that intelligence has a strong
genetic basis. With regard to race differences being
the cause of the observed differences concerned, he
wrote that he thought the issue unproved but that
it seemed probable that genetic factors are involved.
He noted that blacks in Africa had not developed

civilizations comparable to those of Caucasians, Chinese, and other Asians and concluded:

It is entirely possible that the differential performance of cultures must be explained, in part, by differential genetic distribution of aptitudes."[16]

Van den Haag believes that the assertion of most social scientists that there are no genetic racial differences in intelligence is just a contemporary fashion unsupported by convincing evidence. These social scientists, he wrote, "obstinately refuse to act as scientists, being committed to various causes rather than to science." Much of their work can be aptly described as "extremism in the pursuit of egalitarian ideologies."

Thirty years later, van den Haag[17] wrote a comment on Herrnstein and Murray's book *The Bell Curve*.[18] He endorsed their argument that intelligence is important for job success and that there are significant race differences in intelligence. However, he believes they overstated the case for the social importance of intelligence. He suggests that unintelligent youths can attain success as baseball players or as singers of pop songs, while unintelligent girls can become models. He regards their prediction that people with low IQs will become unemployable as "rank speculation" and says that we cannot predict the future. Nevertheless, he commends them for their acceptance of individualism as the paramount value of a liberal society, and once again deplores affirmative action as contrary to this principle.

NOTES

1. van den Haag, E. 1956. *Education as an Industry*. New York, NY: Kelly.

2. Ross, R. & van den Haag, E. 1957. *The Fabric of Society*. New York, NY: Harcourt Brace.

3. van den Haag, E. 1960. Social science testimony in the desegregation cases — a reply to Professor Kenneth Clark. *Villanova Law Review*. 6: 69-79.

4. *Ibid*. 77.

5. van den Haag, E. 1965. Negroes and whites: claims, rights and prospects. *Modern Age*. Fall: 354-362.

6. *Ibid*. 360.

7. *Ibid*. 361.

8. *Ibid*. 362.

9. *Ibid*. 362.

10. *Ibid*. 362.

11. van den Haag, E. 1976. Reverse discrimination. *Review of Education*. September: 427-434.

12. *Ibid*. 429.

13. *Ibid*. 430.

14. van den Haag, E. 1964. *Negroes, Intelligence and Prejudice*. New York: IAAEE Reprint.

15 *Ibid*. 3.

16. *Ibid*. 5.

17. van den Haag, E. 1994. Not hopeless. *National Review*. December 5: 35-36.

18. Herrnstein, R. & Murray, C. 1994. *The Bell Curve*. New York, NY: Free Press.

William B. Shockley, Ph.D.
Stanford University

Chapter 15

William B. Shockley

William B. Shockley (1910–1989), was a physicist and engineer who became one of the founders of the computer electronics revolution with his invention in 1947, with two colleagues, of the junction transistor. This semiconductor device rapidly replaced vacuum tubes in a wide range of electronic devices and soon became a standard component of the new generation of high-tech aircraft, automobiles, calculators, and telecommunications equipment. It was not until he reached his fifties that Shockley became involved in human genetics, race differences in intelligence, and what he called "human quality issues," on which he made important policy proposals.

William Bernard Shockley was born in London in 1910, but shortly after his birth his family moved to California. He was educated at Hollywood High School and the California Institute of Technology, from which he graduated in physics in 1932. He proceeded to the Massachusetts Institute of Technology, from which he obtained his Ph.D. in 1936. He then joined the Bell Telephone Laboratories. In 1941 he was appointed director of research of the U.S. Navy Anti-Submarine Warfare Operations Research Group.

In 1945 Shockley returned to the Bell Telephone Laboratories as director of solid-state physics research and leader of the team which developed the transistor. In 1954, he left Bell Laboratories and founded his own company, Shockley Semiconductor Laboratories, in California's Silicon Valley. From 1958 onwards he taught at Stanford University, at which he was appointed Alexander M. Poniatoff Professor of Engineering Science in 1963. He received numerous awards, including the Medal of Merit for his wartime service, the Maurice Liebman Memorial Prize from the Institute of Radio Engineers, the Oliver E. Buckley Solid State Physics Prize from the American Physical Society and the Cyrus B. Cornstock Award from the National Academy of Sciences. In 1956 he was awarded the Nobel Prize for Physics, jointly with his two colleagues, John Bardeen and Walter Brattain, for the development of the transistor.

It was in the early 1960s that accelerating crime in the United States led Shockley to think about human intelligence, genetics, and eugenics. He said that the single most influential event that led him to ponder these questions occurred in 1963. It was a news story about a teenager, Rudy Hoskins, who blinded a San Francisco delicatessen proprietor by throwing acid at his eyes during the course of a robbery. It transpired that Hoskins was mentally retarded with an IQ of approximately 65, and was one of 17 illegitimate children of a woman with an IQ of 55, who could remember the names of only nine of her children. Shockley believed that this story was symptomatic of the high fertility among the mentally retarded and those with low IQ and that this group would increase in the population by doubling every generation, and eventually would become a majority.[1] These concerns led him to read widely in genetics, psychology and sociology. In 1973 he set up the Foundation for Research and Education in Eugenics and Dysgenics (FREED) as a center for research on these issues. He received financial support from the Pioneer Fund from 1968 through 1978.

It was in 1965 that Shockley first went public with his concerns about population quality in an interview in *US News and World Report*, in which he raised the issues of the high fertility of the less intelligent, the low average IQ of American blacks, the greater dysgenic fertility of blacks as compared with whites, and the social problems arising from these phenomena.[2] The interview was reprinted in the Stanford Medical School alumni magazine, *Stanford M.D.*, as a result of which Shockley was to experience his first attack from the politically correct. This was a letter from the Stanford Genetics Faculty disputing Shockley's views with such words as "pseudoscience" and "malice" and concluding that:

the whole concept of "bad heredity" is in any case a myopic one since the high values of one social milieu are the vices of another one, and our milieu is constantly changing.[3]

The Stanford geneticists' critique of Shockley argued, in effect, that in some "social milieus" the unintelligent are valued more highly than the intelligent, and criminals are more greatly esteemed than the law abiding. They did not specify which societies these are.

INTELLIGENCE AND SOCIAL CAPACITY

Shockley believed that intelligence is important for the maintenance of civilization. He set out this case in a paper presented to the American National Academy of Sciences in 1967.[4] He argued that the intelligence level of a population determined its "social capacity" which he defined as its economic and intellectual achievement and, negatively, its rate of unemployment, crime, and illegitimacy. He illustrated this by presenting a variety of data for blacks and whites in the United States, arguing that an intelligence difference of approximately 18 IQ points between the races could account for the differences between them in these sociological phenomena. Taking data from the 1960s, he derived an index of socioeconomic achievement from the number of Americans in *International Who's Who*. He calculated that American whites were represented in this reference book with a frequency of 1 per 30,000 and blacks with a frequency of 1 per 424,000. A similar differential was present among Americans listed in *American Men of Science*. He showed also that blacks had approximately double the unemployment rate of whites, about 3.5 percent and 7 percent, respectively. Shockley estimated the black crime rate at about 3.5 times greater than the white and illegitimacy rates at about six times greater. He called his paper a "try simplest cases" approach, proposing that the low average black IQ provided the most parsimonious explanation for the low black socioeconomic and intellectual achievement and the high black rates of unemployment, crime, and illegitimacy.

RACE DIFFERENCES IN INTELLIGENCE

Shockley was concerned about the relatively low average IQ of American blacks and the effect of this on the social problems of high unemployment and crime. He followed Henry Garrett and Audrey Shuey in believing that the evidence pointed towards a primarily genetic explanation of the low black IQ, although he thought the evidence was not conclusive, and he repeatedly urged research on this issue. He went public on this in 1968 in a proposal put before the National Academy of Sciences in which he wrote that:

> an objective examination of relevant data leads me inescapably to the opinion that the major deficit in Negro intellectual performance must be primarily of genetic origin and thus relatively irremediable by practical improvements in environment.[5]

To support this conclusion he examined the three leading equalitarian arguments for the low black IQ and cited evidence to show that they were wrong. The first of these was that black IQ was impaired by prenatal or very early environmental disadvantages. Against this, Shockley cited Nancy Bayley's study of 600 black infants who were the equal of 800 white babies in early mental development and ahead of white babies in muscular neurological development. Next, he considered the argument that cultural disadvantage could explain the low black IQ. He countered this by citing evidence that when black children were matched against white for socioeconomic status, the intelligence difference was still present. Finally, he took up the argument that "there is no scientific evidence for racial differences in intelligence." He countered this by the evidence that blacks score relatively better on verbal intelligence tests than they do on non-verbal tests of abstract reasoning.

Shockley's principal contribution to the scientific study of race was to make the case that within the American black population intelligence levels are related to the proportion of white genes, determined by blood groups or skin color.[6] He advanced two arguments in support of this conclusion. First, he cited the blood group evidence collected by T. E. Reed that blacks in Oakland, California, had on average 22 percent Caucasian genes, and those in two counties in Georgia had 11 percent Caucasian genes.[7] He then estimated from military draft data that the Californian blacks had an average IQ of about 90 and those in Georgia of about 80. Thus, he proposed that an extra 11 percent of Caucasian genes in the blacks in California, as compared with those in Georgia, had raised average IQ by about 10 IQ points. He concluded that every 1 percent of Caucasian genes raised the IQ of black-white hybrids by 1 IQ point. He noted that if this rule were extrapolated, blacks with 30 percent of Caucasian genes would have IQs above those of whites. A further extrapolation of the rule implies that blacks with 90 percent of Caucasian genes would have IQs of 160. Shockley realized that there was a problem here and proposed that there must be severe diminishing returns for additional Caucasian genes in blacks after about the 22 percent value typically found among blacks in California.

DYSGENICS

Shockley was disturbed about the dysgenic fertility in the United States, especially in regard to the tendency for higher-IQ couples to have fewer children than unintelligent couples. He made his first public statement on this problem in 1965 at the Nobel Conference on *Genetics and the Future of Man* held at Gustavus Adolphus College, St. Peter, Minnesota.[8] Shockley proposed that the human race was faced with three great threats. These were the threat of nuclear war, the population explosion, and genetic deterioration. He reiterated the views of Galton and the earlier eugenicists that genetic deterioration was taking place in economically developed nations because of relaxation of natural selection by reduced mortality of the less fit, and by dysgenic fertility. Shockley used the noun *dysgenics* for what he called this retrogressive evolution.[9] He suggested that dysgenics was contributing to the increases in crime and illegitimacy which were approximately doubling every decade in the 1950s and 1960s.[10]

Shockley was particularly concerned about the apparently greater dysgenic fertility among blacks than among whites. He cited data from the American census showing that black American married women living with their husbands and aged 35-45 had an average of 4.7 children as compared with 3.8 for comparable white women, indicating greater overall fertility among blacks. However, among the wives of professional or technical workers with at least one year of college education, blacks had on average 1.9 children as against 2.4 for whites.[11] Later he cited some of the results of the 1970 American census which confirmed this trend in showing that black farm women aged 35-44 had on average in the United States in the 1950s and 1960s 5.4 children as compared with 1.9 for black women college graduates, while for whites the two averages were 3.7 and 2.3 respectively.[12]

Shockley believed that this must mean that the intelligence of blacks was deteriorating relative to that of whites. He cited evidence suggesting that this was indeed the case. He showed that the World War I draft data showed a black "overlap" of 13 percent, i.e. 13 percent of blacks scored above the white median, whereas the Surgeon General's military data for 1966 showed an overlap of only 7 percent. He estimated that this indicated a deterioration in the average IQ of American blacks of 5 IQ points over the course of approximately two generations from 1918 to 1966. He believed that if this trend continued it would lead to the "genetic enslavement" of American blacks, that is to say that increasing numbers of them would become mentally retarded and unable to find a role in society.[13] He thought that blacks were already at a disadvantage in the United States by virtue of their lower average IQ and that this disadvantage would increase as the intelligence levels fell further. He called this "the tragedy of the American Negro."[14]

EUGENIC POLICY PROPOSALS

At various times in his writings Shockley made or supported five proposals for policies to counteract dysgenics. He first set some of these out in 1965 and elaborated on them on a number of subsequent occasions.[15] His first proposal was his endorsement of the scheme first suggested by Hermann Muller for the establishment of elite sperm banks which women would be encouraged to use. Such a bank was set up in California around 1970 and stored the sperm of three Nobel laureates and several other eminent persons, and for which Shockley was himself a donor.

Second, Shockley approved a suggestion made by Ernst Mayr, the noted Harvard zoologist, that tax allowances for children should be given as a percentage of income. The effect of this would be that high earners, who may be presumed to have genetically desirable qualities, would gain greater tax relief than low earners, and hence have an incentive to have more children than they would otherwise plan.

Third, Shockley advocated that abortion should be legalized for unmarried women and for those who might have reason to believe that they would produce a genetically defective infant. He did not specify who these might be, but probably had in mind women who were mentally retarded or possible carriers of harmful dominant genes.

Fourth, Shockley favored the more extensive sterilization of the mentally retarded. He noted that legislation providing for this was in place in a number of states of the U.S., but that the number of sterilizations carried out in the post World War II years had greatly declined. In 1968 he proposed that "possibly" there should be mandatory sterilization of women "after the nth successive illegitimate child on relief with n to be determined by national vote and possibly constitutional amendment."[16]

In 1972 Shockley put forward an even more radical sterilization proposal as "a thinking exercise."[17] This was that all non-taxpayers with IQs below 100 would be offered a payment to be sterilized. The payment would be at the rate of $1,000 for each IQ point below 100, so that, for instance, someone with an IQ of 80 would be paid $20,000. This sum would be put in a trust fund for the individual concerned. He thought that the public savings arising from not having to maintain retarded children would greatly outweigh the costs involved. The stipulation that the sterilization bonuses would only be paid to those not paying income tax would rule out the great majority of the population and confine the plan to low earners and the unemployed. On the other hand, the bonus would also be available to those with certain genetic diseases, including diabetes, epilepsy, and arthritis. The proposal was an interesting exercise in voluntary eugenics although the details needed further specification. For example, applicants who already had all the children they wanted might give up employment in order to qualify and then fake an IQ of zero. This would entitle them to a payment of $100,000, a considerable sum in the early 1970s. There might be millions of applicants for these bonuses which would put an intolerable strain on the national finances.

Shockley's fifth contribution to practical eugenics was his support for a program proposed in the 1960s by the economist Kenneth Boulding. The program was designed to produce a numerically stable population. The proposal was that every girl at the start of puberty would be fitted with a long-acting contraceptive. She would also be given a certificate entitling her to have 2.2. children. When she wanted to have a baby, she would have to surrender her certificate to an authorized physician, who would remove the contraceptive. On the birth of her baby the contraceptive would be replaced, and she would be given a new certificate entitling her to have 1.2 babies. The same procedure would be followed if she wished to have a second baby. A further feature of the proposal was that these baby entitlement certificates could be bought and sold. Once a woman had had two babies she could either sell her remaining allocation of 0.2 babies or buy another 0.8 units to obtain a 1.0 certificate, which would entitle her to have another baby.

Boulding's proposal was put forward as a population control mechanism devised to stabilize population or limit it to a small increase. However, Shockley welcomed it because he thought it would have a eugenic effect. What would happen, he thought, would be that poor and typically less intelligent women would sell their baby entitlement certificates and better off women would buy them with the result that the fertility of the poor and unintelligent would decline and that of the wealthy and more intelligent would increase. In addition the plan would have a eugenic impact because the low intelligence women would have to obtain a certificate to have more than two children and have their contraceptive device removed. Far more low intelligence than high intelligence women have children by accident, and since the effect of the plan would be to eliminate accidental births after the first two, the effect would be eugenic. It has to be admitted, however, that it would be difficult to enact the proposal in a democracy.

Shockley expounded his views on the seriousness of genetic deterioration, the genetic basis of the low black IQ, and the need for eugenic measures in numerous radio and TV appearances and in a number of public lectures. At many of these he was shouted down by hostile members of the audience. In 1972 his image was burned in effigy at a demonstration on the Stanford campus. In 1974 he was humiliated by the University of Leeds,

England, which was about to bestow an honorary degree on him but withdrew at the last minute when the authorities were scared off by his conclusions of race differences in intelligence. It was typical of Shockley that he promptly issued a press release to get maximum publicity for his views. In 1981 he sued the Atlanta Constitution for $1 million over an article which compared his eugenic theories to those the Nazis. He won the case but was awarded only one dollar in damages and had to pay his own legal costs.

Shockley was a courageous and tireless campaigner for research into the causes of human and race differences and for thoughtful consideration of eugenics. He died on 12 August 1989, at the age of 79.

NOTES

1. Shockley, W. B. 1969. Human quality problems and research taboos. *New Concepts and Directions in Education*. 28: 67-99.

2. Shockley, W. B. 1965. Is quality of U.S. population declining? *U.S. News and World Report*. 22 November.

3. Faculty of the Department of Genetics. 1966. The issue of bad heredity. *Stanford M.D.* Series 5, No. 2.

4. Shockley, W. B. 1967. A try simplest cases approach to the heredity-poverty-crime problem. *Proceedings of the National Academy of Sciences*. 57: 1767-1774.

5. Shockley, W. B. 1968. Proposed research to reduce racial aspects of the environment-heredity uncertainty. *Science*. 160: 443.

6. Shockley, W. B. 1971. Hardy-Weinberg law generalized to estimate hybrid variance for Negro populations and reduce racial aspects of the environment-heredity uncertainty. *Proceedings of the National Academy of Sciences*. 68: 1390.

7. Reed, T. E. 1969. Caucasian genes in American Negroes. *Science*. 165: 762-768.

8. Shockley, W. B. 1965. Population control or eugenics. In J. D. Roslansky (ed.) *Genetics and the Future of Man*. Amsterdam: North-Holland.

9. Shockley, W. B. 1972. Dysgenics, geneticity, raceology. *Phi Delta Kappan*. Supplement. 297-312.

10. Shockley, W. B. 1969. Offset analysis description of racial differences. *Proceedings of the National Academy of Sciences*. 64: 1432.

11. Shockley, W. B. 1972. *Op. cit.*

12. Shockley, W. B. 1975. True (not berserk) humanitarianism: A positive absolute value that unites religion and science. *Proceedings of the Fourth International Conference on the Unity of the Sciences*. New York.

13. Shockley, W. B. 1967. *Op. cit.*

14. Shockley, W. B. 1981. Intelligence in trouble. *Leaders Magazine*. April: 229-233.

15. Shockley, W. B. 1965. *Op. cit.*

16. Shockley, W. B. 1968. Proposed research to reduce racial aspects of the environment-heredity uncertainty. *Paper presented to the National Academy of Sciences.* 24 April.

17. Shockley, W. B. 1972. *Op. cit.*

PART III:
The Nature of Intelligence

Philip E. Vernon, Ph.D.
University of London

Chapter 16

Philip E. Vernon

Philip E. Vernon (1905–1987) was one of the foremost authorities on intelligence and personality during the middle decades of the 20th century. He wrote many books and papers on these subjects over the course of more than half a century. His scholarship and judicious treatment of controversial issues have been widely acknowledged and his conclusions commanded the respect even of those who disagreed with him.

Philip Ewart Vernon was born in 1905 in Oxford, England, the only son of H. M. Vernon, a lecturer in physiology at the University of Oxford and the author of several books on industrial psychology. It was he who encouraged both his son, Philip, and his daughter, Magdalene, to study psychology. Both children followed his advice, and Magdalene eventually became professor of psychology at the University of Reading.

Philip E. Vernon was educated at Oundle, a British public school with a strong reputation in science, and at the University of Cambridge, where he first studied physics, chemistry, and physiology, followed by specialization in psychology. After graduating, Vernon remained at Cambridge and worked on the psychology of music for his Ph.D. In 1927 he was awarded the Laura Spelman Rockefeller Fellowship. This enabled him to travel to the United States to work on personality testing with Mark May at Yale, and then with Gordon Allport at Harvard, with whom he wrote *Studies of Value* [1] and *Studies in Expressive Movement*.[2]

Vernon returned to England in 1931 and took up a research and teaching fellowship at Cambridge, and then moved to London to work at the child guidance clinic at the Maudsley Hospital. In 1938 he accepted a position as head of the psychology department at the University of Glasgow, and in the same year published two books, *The Assessment of Psychological Qualities by Verbal Methods*[3] and *The Standardization of a Graded Word Reading Test*.[4] In 1940 he was appointed psychological research advisor to the British Admiralty and War Office. Here he worked on personnel selection tests, on which he wrote a book in collaboration with J. B. Parry.[5] In 1946 Vernon was appointed professor of educational psychology at the University of London Institute of Education. He held this position for 22 years, during much of which he acted as editor of the *British Journal of Educational Psychology*. In 1968 he moved to Canada to become professor of educational psychology at the University of Calgary, primarily to get away from the damp British winters for reasons of health.

TEXTBOOKS ON HUMAN ABILITIES

Vernon wrote three general textbooks on intelligence: *The Measurement of Abilities*,[6] *The Structure of Human Abilities*,[7] and *Intelligence: Heredity and Environment*.[8] His approach was in the tradition of the London School of differential psychology pioneered by Charles Spearman and continued by Sir Cyril Burt and Raymond Cattell. This used factor analysis to demonstrate the existence of a general factor, designated Spearman's g, which affects the performance of all cognitive tasks, and a number of so-called specific factors that affect performance on particular tests. In the 1930s this theory was opposed by the American psychologist, L. L. Thurstone. Thurstone denied the existence of g and instead posited six major independent cognitive factors which he termed the Primary Mental Abilities: reasoning (R), verbal comprehension (V), numerical ability (N), spatial ability (S), perceptual speed (P), and word fluency (W).

In the late 1940s Vernon and Burt both worked on these alternative theories and came up with the same solution. This consisted of the hierarchical model of intelligence. In this Spearman's g (general intelligence) stands at the apex of a pyramid, and subdivides into two major group factors, designated: *v:ed* (verbal-numerical-educational abilities) and *k:m* (practical-mechanical-spatial-physical abilities). These are in turn broken down into primary abilities, and below them into narrow specific factors. Burt published his hierarchical model in 1949[9] and Vernon a year later, so Burt has to be accorded priority. Vernon did not cite Burt's work in his 1950 book, so he must have arrived at the same conclusion independently. This model has become widely although not universally accepted as the leading theory of the structure of intelligence in the 1980s and 1990s.

Vernon's third textbook, *Intelligence: Heredity and Environment*,[10] was a thorough and impartial account of what was known about intelligence and the evidence for and against the opposing opinions on controversial topics such as the role of heredity and environment and race differences. Vernon concluded that approximately 60 percent of individual differences in intelligence are attributable to genetic factors. He also concluded that there is some genetic basis to racial differences in intelligence, writing that:

> In order to account for the observed differences between the races, it would still seem logical to include some genetic component as well as environmental differences. [11]

This textbook was widely used in college courses in the 1980s.

SOCIAL CLASS DIFFERENCES IN INTELLIGENCE

Numerous studies have found that there are substantial social class differences in intelligence. For instance, the analysis of the World War I American military conscripts showed that the average IQ of socioeconomic class 1 (professional) was 123 and that average IQs declined with social class to 96 in socioeconomic class 5 (unskilled workers).[12] The hereditarian position is that these social class differences have a genetic basis. They arise because over the course of many generations those with higher intelligence have tended to rise in the social class hierarchy while those with lower intelligence tend to fall (social mobility). People tend to marry those who are like themselves (assortative mating), and the two parents transmit their genes to their children. In this way the genetically based social class hierarchy becomes established. It is for this reason that eugenicists became concerned about the relatively low fertility of the professional class that began to appear in North America and Europe in the middle decades of the 19th century.

The view that the social class differences in intelligence have a genetic basis has frequently been disputed by sociologists and equalitarians. For instance, Richard Lewontin has asserted that "There is not an iota of evidence that social classes differ in any way in their genes."[13] Vernon came down on the hereditarian side of this issue. After reviewing the evidence, he asserted in *Intelligence and Cultural Environment* that "There are certainly some genetic differences between the classes,"[14] and later he endorsed Sir Cyril Burt's conclusion that:

> Assortative mating and social mobility lead to an equilibrium of genetic differences among upper, middle and lower classes.[15]

In support of this verdict, Vernon cited evidence that the intelligence of adopted children is positively related to the social class of their natural fathers. Curiously, Lewontin did not discuss this evidence in making his assertion that there is no genetic basis to social class differences in intelligence.

RACE DIFFERENCES IN INTELLIGENCE

Vernon tackled the issue of race differences in greater detail in 1969 with his book *Intelligence and Cultural Environment*.[16] This was a review of previous studies, supplemented by an account of his own work testing the IQs of 11 year old boys in England, Scotland, Jamaica, Uganda, and Canada (where he tested Canadian Indians and Eskimos). Vernon administered 16 tests to samples of 50 boys in each of these countries. He did not calculate the average IQs of his samples. He preferred to report the medians of each of the tests. In general these showed average IQs of around 75 in Jamaica and Uganda, 80 for Canadian Indians, and 92 for Eskimos. Vernon's interpretation of these results was that they reflected both genetic and environmental factors.

Vernon accepted the fact that some of these peoples lived in impoverished conditions and were probably adversely affected by disease and malnutrition. Nevertheless, he wrote that:

> To a very great extent man makes himself and fashions his own environment, and it is he who must be changed if he is to achieve a more prosperous and healthy existence. [17]

This was an anticipation of what was later to be called the principle of genotype-environment correlation which states each individual, to some extent, constructs his own environment. However, Vernon believed that the large differences in average intelligence levels found between different races in different cultures, as for instance the low average IQs found in Africa and the Caribbean, are largely environmentally determined. On this issue he concluded that:

> When environmental differences are more extreme, as between ethnic groups, their effects predominate. This does not mean that there are no innate differences in abilities, but they are probably small and we have no means of proving them.[18]

THE INTELLIGENCE OF EAST ASIAN AMERICANS

Vernon's last book was his 1982 study of the intelligence of Americans of East Asian origin, largely ethnic Chinese and Japanese.[19] His work for this book was supported by the Pioneer Fund. Vernon's interest in this subject was aroused by

The Science of Human Diversity
A History of the Pioneer Fund

Richard Lynn's 1977 studies showing that the Japanese in Japan and the Chinese in Singapore obtained higher average IQs than whites in North America and Europe, at approximately 106 and 110, respectively, and that they scored particularly high on tests of spatial ability. Vernon examined the evidence and found that East Asian Americans obtain mean scores of approximately 110 on non-verbal reasoning and spatial ability and 97 on verbal ability. These can be averaged to give a general IQ of 106.

This conclusion has been challenged by James Flynn who estimates the IQ of East Asian Americans at approximately 99.5 for what he calls "non-verbal ability" and 95.5 for verbal ability, thus giving an average of approximately 97.5.[20] However, these estimates omit spatial ability, in which East Asians are particularly strong, so they inevitably bias the result downwards. The more recent results of the standardization of the Differential Ability Test in the United States give the ethnic Asians a mean IQ of 107, closely confirming Vernon's estimates.[21]

Vernon was by temperament the model cautious scientist who sees all sides of a controversy and is unwilling to go beyond the evidence. He was a meticulous scholar who was at his best consolidating and synthesizing the work of others. When he did draw conclusions, such as that intelligence has a high heritability and that there is some genetic difference between the races, they had high credibility. His readers could be confident that if even Vernon had endorsed a conclusion, the evidence for it must be overwhelming.

Vernon received a number of honors. He was President of the Psychology Section of the British Association for the Advancement of Science (1952); President of the British Psychological Society (1945-55) and Honorary Life Member of this Society; President of the Industrial, Educational, and Scottish Sections of the British Psychological Society, and Chairman of the Society's Committee on Secondary School Selection; Vice President of the Eugenics Society (1961); Honorary Fellow of the International Association for Cross-Cultural Psychology; Fellow of the American Psychological Association; Life Fellow of the Canadian Psychological Association; Fellow of the U.S. National Academy of Education (1968); Fellow of the Center for Advanced Study in the Behavioral Sciences (1961 and 1975); Fellow of the Scottish Council for Research in Education; and first recipient of the Distinguished Lecturer Award of the Faculty of Education at the University of Calgary. Philip Vernon died in Calgary, Canada, in 1987.

NOTES

1. Allport, G. W. & Vernon, P. E. 1931. *Studies of Value: A Scale for Measuring the Dominant Interests in Personality.* Boston, MA: Houghton Mifflin.

2. Allport, G. W. & Vernon, P. E. 1932. *Studies in Expressive Movement.* New York, NY: Macmillan.

3. Vernon, P. E. 1938. *The Assessment of Psychological Qualities by Verbal Methods.* London, U.K.: Her Majesty's Stationery Office.

4. Vernon, P. E. 1938. *The Standardization of a Graded Word Reading Test.* London, U.K.: University of London Press.

5. Parry, J. B. & Vernon, P. E. *Personnel Selection in the British Forces.* London, U.K.: University of London Press.

6. Vernon, P. E. 1940. *The Measurement of Human Abilities.* London, U.K.: University of London Press.

7. Vernon, P. E. 1950. *The Structure of Human Abilities.* London, U.K.: University of London Press.

8. Vernon, P. E. 1979. *Intelligence: Heredity and Environment.* San Francisco, CA: W. H. Freeman.

9. Burt, C. L. 1949. The structure of the mind: a review of the results of factor analysis. *British Journal of Educational Psychology.* 19: 100-111.

10. Vernon, P. E. 1979. *Op. cit.*

11. *Ibid.* 267.

12. Johnson, D. M. 1948. Applications of the standard IQ score to social statistics. *Journal of Social Psychology.* 27: 217-227.

13. Lewontin, R. 1993. *The Doctrine of DNA.* London, U.K.: Penguin. 23.

14. Vernon, P. E. 1969. *Intelligence and Cultural Environment.* 67.

15. Vernon, P. E. 1979. *Op. cit.* 250.

16. Vernon, P. E. 1969. *Op. cit.*

17. *Ibid.* 214.

18. *Ibid.* 215.

19. Vernon, P. E. 1982. *The Abilities and Achievements of Orientals in North America.* New York, NY: Academic Press.

20. Flynn, J. R. 1991. *Asian Americans: Achievement beyond IQ*. Hillsdale, NJ: Erlbaum.

21. Lynn, R. 1996. Racial and ethnic differences in the United States on the differential ability scale. *Personality and Individual Differences*. 20: 271-273.

Arthur R. Jensen, Ph.D.
University of California at Berkeley

Chapter 17

Arthur R. Jensen

Arthur R. Jensen (born 1924) has been a prolific and influential researcher and writer on intelligence from the 1960s until the present. He is the leading exponent of the position that intelligence is a unitary entity, an important factor in educational and occupational achievement, and that there is some genetic basis to social class and race differences in intelligence.

Arthur Robert Jensen was born in 1924 in San Diego, California. His paternal grandparents were immigrants from Copenhagen, Denmark, his maternal grandparents from Berlin, Germany. Jensen's father owned a lumber and building supplies business in San Diego, which provided a comfortable living for his family. As an adolescent Jensen was absorbed by classical music, in which he intended to make his career. However, he realized that he lacked the talent to get to the top of the music profession, so he abandoned his hopes for a musical career and entered the Berkeley campus of the University of California, where he majored in psychology. After graduation, he worked at a variety of jobs including his father's lumber business, as a technician in a pharmaceutical laboratory, as a social worker, as a high school biology teacher and orchestra conductor, and as a preparator in the Zoology Department of Columbia University in New York City.

In the late 1940s Jensen returned to San Diego to teach in high school and decided to pursue psychology seriously. He took a master's degree at the San Diego State University in 1950 and entered the Ph.D. program at Columbia University, where he worked on the Thematic Apperception Test as a measure of aggression. In 1955, after obtaining his Ph.D., Jensen served a year as a clinical intern in Baltimore. He spent the next two years, 1956-1958, working with Hans Eysenck at the Institute of Psychiatry in London. In 1958 he returned to the University of California, Berkeley, as assistant professor of educational psychology in the Graduate School of Education. He became associate professor in 1962, and full professor in 1966. He received a Guggenheim Fellowship in 1964, which enabled him to spend another year in London, to work in Eysenck's department. Jensen's work on intelligence has been supported by the Pioneer Fund from 1973 onwards.

LEVEL I AND LEVEL II ABILITIES

During the 1960s Jensen worked on the learning abilities and intelligence of mentally retarded children with IQs below 75. He noticed that a number of Mexican-American children placed in classes for the "educable mentally retarded" (EMR) seemed more "normal" in their play and their perceptual-motor capacities than white middle class children diagnosed as educable mentally retarded. This led him to investigate systematically the learning abilities and intelligence of mentally retarded children in relation to their ethnicity, race, and social class. The results of this work were that minority children in EMR class performed, on average, better on the learning tests than middle-class white children of the same IQ. Many of the minority children classed as EMR performed on the learning tests within the range of most of the middle-class white children in regular classes. However, the learning tests, although moderately correlated with IQ, were much poorer predictors of scholastic performance than the IQ. Moreover, the learning tests had significantly lower correlations with IQ in the minority children than in the white majority. It was not unusual to find minority children in retarded classes who performed at an average or above average level on the learning tests, but it was rare to find white children in these classes who did so.

To make theoretical sense of these results, Jensen hypothesized that two kinds of ability are involved, which he termed Level I (measured by the direct learning tests) and Level II (measured by IQ). Level I ability requires only the accurate registration of the information input, such as the digit span test, which requires the accurate immediate recall of several numbers. Level II ability requires that some mental transformation or manipulation be performed on the information input in order to arrive at the correct output.

Intelligence tests are largely measures of Level II abilities, although some tests contain one or two Level I abilities as well. Jensen advanced the hypothesis that individuals differed on these two kinds of ability in such a way that the scatter diagram of the relation between Levels I and II formed what is known as a "twisted-pear" correlation, such that low Level I ability dependably predicted low Level II ability, but high Level I ability had little predictive relationship to Level II ability. Jensen found that among white children, socioeconomic status differences were greater on Level II than on Level I ability.[1] He also found that while black and Hispanic children differ considerably less from white children on Level I ability than on Level II ability,[2] the diagnostic criterion for mental retardation in the schools was based almost entirely on Level II ability. The failure to recognize the majority-minority difference in Level I ability among those children placed in retarded classes, therefore, meant that the usual instructional methods failed to capitalize on the normal Level I ability possessed by many of the minority children. Despite low IQ, their capacity for learning on a Level I basis was better than that of most of the white children in EMR classes.

Thus Jensen's research turned to the scholastic problems of minority children, particularly the cultural and environmental factors that shape their distinctive patterns of abilities. He conjectured that special methods of instruction could bring their Level I ability to bear on learning the basic school subjects. Jensen's experiments on the verbal mediation of learning and problem solving showed that it improved performance but that educationally retarded children had much less inclination spontaneously to verbalize the elements of the task than did educationally successful children. When the retarded children were explicitly instructed to verbalize aloud about the goal of the task and the essential elements of the problem, their learning was markedly improved. In some cases they even came up to the level of pupils with average IQs, who engaged in similar verbalization implicitly and spontaneously, without having to be reminded on each occasion to do so. This tendency spontaneously to verbally mediate learning therefore appeared to be only an epiphenomenon of Level II ability rather than the cause of it. Although verbal mediation could be trained up for a given task, in most EMR pupils it did not transfer spontaneously to tasks that differed from the specific training task.

The Harvard Educational Review Article

In the 1960s Jensen shared the prevailing view among social scientists that intelligence was largely or entirely environmentally determined. According to this view, white middle class children benefited from a number of environmental advantages which fostered their intelligence and led them to perform well on intelligence tests.

Conversely, lower class white children and black and Hispanic children were relatively deprived of these environmental advantages, and this retarded the development of their intelligence. In 1966-1967 Jensen obtained a Guggenheim fellowship to spend a sabbatical year at the Center for Advanced Study in the Behavioral Sciences, and he planned to spend this time writing a book on the adverse effects of cultural disadvantages on the intelligence and educational attainment of American ethnic and racial minorities. As he read the literature on this issue Jensen realized that the genetic aspect of mental ability had been neglected or even misrepresented in textbooks on intelligence. Instead of writing a book, Jensen wrote a long article for the *Harvard Educational Review* in which he set forth the evidence for the existence of general mental ability (known as the *g* factor), the reliability and practical validity of intelligence tests, the question of cultural bias of IQ tests for minorities, the heritability of intelligence, the genetic and environmental factors responsible for social-class and racial differences in IQ, and the proper uses of tests in education and employment.[3]

With regard to social class differences in intelligence, Jensen concluded that through the process of social mobility over the course of generations, more intelligent people tend to rise in the social hierarchy, and hence the genes for intelligence come to be to some degree segregated by social class. With regard to the different average IQ between blacks and whites, Jensen concluded that:

> The preponderance of the evidence is, in my opinion, less consistent with a strictly environmental hypothesis than with a genetic hypothesis, which, of course, does not exclude the influence of environment or its interaction with genetic factors.[4]

Jensen's article received extensive coverage in the media, which focused especially on his conclusion that the evidence pointed to some degree of genetic involvement in the black-white average IQ difference. Radical groups on the Berkeley campus were outraged and demonstrated at Jensen's lectures and called for his dismissal.

RACE DIFFERENCE IN INTELLIGENCE

Jensen spent the early 1970s defending and amplifying his conclusion that the research evidence points to a genetic basis for part of the difference in average IQs between American blacks and whites. This work appeared in two books, *Genetics and Education*[5] and *Educability and Group Differences*.[6] The general position argued by Jensen in these two books is that the equalitarian explanations for the black-white difference do not withstand critical examination and that a number of converging lines of evidence point to a primarily genetic explanation of the black-white difference, none of which is in itself conclusive but which taken together add up to a strong genetic case.

Jensen examined in detail the argument of equalitarians that the low average black IQ is the result of the lower socioeconomic status of blacks.

He found that even when blacks and whites are matched for socioeconomic status, there is still a difference of around 12 IQ points between them. Furthermore, he pointed out that Native American Indians have lower average socioeconomic status than blacks, yet achieve higher average IQs. He also addressed a number of other common equalitarian arguments. First, he showed that intelligence tests are not culturally biased against blacks. Second, the lower average IQ of blacks cannot be explained in terms of formal educational disadvantages because it is fully present among preschool children. Third, compensatory education, such as Head Start and similar programs, has done little to raise the IQs of blacks, although it sometimes increases specific cognitive skills for a limited period. Fourth, the hypothesis that the low average black IQ and educational performance is caused by low teacher expectations is not supported. Fifth, a variety of other equalitarian explanations of the low black IQ do not stand up to critical examination. These include low motivation while taking the test, the administration of the test by white psychologists, malnutrition, verbal deprivation in childhood, and styles of child rearing.

Jensen's most subtle argument for a genetic component to the black-white IQ difference is that equalitarians have not been able to identify the environmental factors accounting for substantial differences in intelligence between blacks and whites in circumstances where the heritability of intelligence is high in both races. The high heritability implies that the putative environmental factor depressing intelligence in blacks must be largely confined to the black population and also be present among virtually all blacks. The factor must be largely confined to the black population because if it were also present in some of the white population, the white heritability would be lower; and it must be present in virtually all blacks because otherwise the heritability among blacks would be lower. It has not proved possible to find a plausible environmental factor which could fulfill these conditions.

Suppose we were to postulate that the factor depressing the black IQ is the absence of vitamin X, a nutrient which is essential for the development of intelligence. It is impossible to imagine that an essential nutrient could be present in the diet of virtually all whites and absent from the diet of virtually all blacks in a society like contemporary America where everyone buys broadly the same kinds of foods from the same sorts of stores. The same problem is present with all of the environmental factors which have been advanced to explain the low black IQ such as inferior schools, one-parent families and low incomes. None of these is present for virtually all blacks and for very few whites.

In 1973 Jensen reaffirmed in his *Educability and Group Differences* the conclusion he had reached in his 1969 *Harvard Educational Review* article, that:

> in view of all the most relevant evidence which I
> have examined, the most tenable hypothesis, in my

judgment, is that genetic, as well as environmental, differences are involved in the average disparity between American Negroes and whites in intelligence and educability, as here defined. All the major facts would seem to be comprehended quite well by the hypothesis that something between one-half and three-fourths of the average IQ difference between American Negroes and whites is attributable to genetic factors, and the remainder to environmental factors and their interaction with genetic differences.[7]

BIAS IN MENTAL TESTING

One of the most persistent criticisms of intelligence tests has been that they are biased in favor of the white middle class psychologists who construct them and against lower class whites and ethnic and racial minorities. In the 1970s Jensen devoted much of his time to an examination of this thesis. The result was the publication of his *Bias in Mental Testing*,[8] an 800-page exhaustive treatment of the problem. Jensen concluded that the test bias argument does not hold up and summarized his views as follows:

Most current standardized tests of mental ability yield unbiased measures for all native-born English-speaking segments of American society today, regardless of their sex or their racial and social-class background. The observed mean differences in test scores between various groups are generally not an artifact of the tests themselves, but are attributable to factors that are causally independent of the tests.... Whatever may be the causes of group differences that remain after test bias is eliminated, the practical applications of sound psychometrics can help to reinforce the democratic ideal of treating every person according to the person's individual characteristics, rather than according to his or her sex, race, social class, religion, or national origin.[9]

Following the success of *Bias in Mental Testing*, Jensen's publisher urged him to do for the lay public a relatively short book touching on all the main topics discussed in his previous works, but with far less technical material. So Jensen wrote *Straight Talk About Mental Tests*, a highly readable overview of his position on mental abilities, tests, heredity, environment, social class, race, and education.[10]

THE NATURE OF SPEARMAN'S *g*

In 1904 the British psychologist Charles Spearman advanced the theory of the existence of a unitary mental power which partially determines the efficiency with which all mental tasks are performed.[11] Spearman called this unitary power general intelligence or *g* and proposed that it was responsible for the positive correlations always observed between different cognitive tests. From the early 1980s Jensen concentrated much of his research efforts on the elucidation of the nature of *g*. He revived Galton's theory that the underlying basis of general mental ability might be mental speed and carried out a program of systematic research on individual differences in the speed and efficiency of information processing, as measured by reaction time (RT) in a variety of elementary cognitive tasks (ECTs), and their relation to intelligence as measured by intelligence tests. Jensen quickly discovered that replicating Galton's method for measuring RT, based on only a few trials for each subject, yielded individual measures with exceedingly low reliability coefficients, only .10 to .20, which largely explained Galton's unpromising results. However, by increasing the number of test trials and using an electronic apparatus that measured reaction times accurately to the millisecond, Jensen obtained reliability coefficients above .90.

Proper testing procedures and advances in statistical methods since Galton's day now permit definitive analyses of the speed of mental processes and their relation to individual differences on standard intelligence tests. Whereas a person's answers to items on ordinary IQ tests can reveal only the end result of the brain's information-processing activity, the person's reaction times measure the speed and efficiency of operation of the basic information processes themselves. Compared to the items in typical IQ tests, the reaction time tasks are so simple that everyone can do them correctly, typically in less than one second. Reliable individual differences show up only in the speed and consistency of response. For example, the person looks at an array of one, two or more light bulbs while pressing a "home" button; when one of the lights goes "on" the person releases the home button as quickly as possible and presses another button, located a few inches from the home button, which turns the light off.

In this experimental procedure, reaction time (RT) is defined as the time taken to release the home button when the light goes on, and movement time (MT) is the time between releasing the home button and pressing the button that turns the light off. Reaction time increases with the number of light bulbs in the array and typically ranges between 200 and 500 milliseconds. Movement time is considerably shorter than reaction time and is unrelated to variation in the cognitive processing demands of the task, being fairly constant across differing amounts of information conveyed by the reaction stimulus. Jensen found that individual difference in RT (and to a lesser degree, MT) were correlated with scores on IQ tests given without time limit. More interesting was the finding that individual variability (i.e. inconsistency) in RT across a number of trials is negatively correlated with IQ even more strongly than is the person's average RT. Jensen viewed this consistency measure (i.e., the standard deviation of the subject's RTs) as an indicator of the efficiency of the individual's information processing system, which, in addition to the average speed of its operations, is the basis of psychometric g.[12]

Jensen has used a number of different RT tasks to measure the speed and efficiency of the several distinct information processes hypothesized to be involved in g, such as simple stimulus apprehension, discrimination, choice, and retrieval of the information from short term and from long term memory. He found that RT provided a sensitive measure of the cognitive processing system, yielding measures which, in combination, correlate .60 to .70 with IQ, consistent with Galton's hypothesis that the speed and efficiency of information processing are the basis of individual differences in general mental ability or *g*.

Besides RT measurements, Jensen also used an electronic tachistoscope to measure "inspection time" (IT), the duration of stimulus presentation needed by the subject to make a very simple visual discrimination with a specified probability of being correct, such as judging which of two closely parallel lines is longer when one line is twice as long as the other. This IT measure, which averages only 40 to 50 milliseconds in college students, correlates about .50 with IQ.

THE NEUROPHYSIOLOGICAL BASIS OF INTELLIGENCE

To probe the neural basis of individual differences in mental processing speed, Jensen collaborated with experts in the use of electroencephalographic techniques to measure such variables as the latency and amplitude of the average evoked potential, i.e. the speed and magnitude of the brain's electrical response to an auditory stimulus, and neural conduction velocity in the visual path from the retina to the visual cortex. He and T. E. Reed found that all these measures are significantly correlated with IQ, which argues for a fundamental neurophysiological basis of intelligence in the speed and accuracy of neural efficiency.[13]

PHYSIOLOGICAL CORRELATES OF SPEARMAN'S *g*

Not all intelligence tests are equally good measures of *g*. In general, tests of problem solving or reasoning are the best measures of *g*, while tests of simpler cognitive processes, such as short term memory, are poorer measures of *g*. In the early 1990s Jensen gathered evidence from the world literature on every known physical correlate of *g* and of the degree to which various tests are measures of *g*. He found that the extent to which tests are measures of *g* is directly related to their heritability coefficient, as determined from twin studies, their correlation with the brain's evoked electrical potentials, the degree to which test scores are lowered by the genetic effect of inbreeding (as in the offspring of cousin mating), and the tests' correlations with head and brain size. These findings indicate that the *g* factor, more than any other factor measured by mental tests, reflects the biological basis of individual differences in mental ability.[14]

In an investigation of the relationship between head size (taken as a proxy for brain size) and IQ, Jensen showed that the correlation between head size and IQ exists within families and is therefore most probably a pleiotropic relationship, i.e., variation in both traits is caused by the same genes. The analytic advantage of a within-family correlation is that it directly controls for all of the known as well as the unknown environmental, socioeconomic, and cultural differences that exist between families in the population. The resulting within-family correlation, therefore, represents an intrinsic or functional relationship, in the sense that it is not simply an incidental result of population heterogeneity in the genetic or environmental basis of each of the correlated traits.[15] Again using the within family methodology in comparing high IQ students with their lower IQ full siblings, Jensen discovered a positive within-family correlation, consistent with pleiotropy, between IQ and degree of myopia measured as a continuous variable. He argued that the existence of the many physical correlates of intelligence, some of them pleiotropic, means that IQ measures something more than just a superficial cultural artifact. The evidence indicates that g has a neurophysiological basis and must have evolved during the evolution of *Homo sapiens*.

BLACK-WHITE DIFFERENCES IN SPEARMAN'S g

In the early 1980s Jensen returned to the problem of the nature of the black-white difference in intelligence which he had conceptualized in the 1960s in terms of Level I and Level II abilities. His new approach was to take up a hypothesis advanced by Spearman in 1927 to the effect that the size of the mean black-white differences on various tests reflects the differences in the extent to which the tests are measures of g.

Jensen investigated Spearman's hypothesis in 11 independent data sets in which a large variety of mental tests were obtained on representative samples of blacks and whites. He found that in every one of the data sets, Spearman's hypothesis was borne out.[16] What Spearman's hypothesis means is that the main factor in the black-white difference in cognitive abilities is not the result of any particular type of test-item contents or specific knowledge or skills, but is mainly a function of g, the one factor common to all cognitive tests regardless of their specific item contents. The crucial result is that the more that a test measures the g factor, the larger is the mean black-white difference on the test. Jensen has also demonstrated that the g factor is present in every large battery of diverse tests examined so far and that one and the same g factor emerges for both blacks and whites.

In a follow-up study on this theory, Jensen carried out a critical test of Spearman's hypothesis, using 24 RT and MT measurements of speed and efficiency of information processing in a battery of elementary cognitive tasks (ECTs) that bore no resemblance to conventional cognitive tests. The ECTs were so simple that the average response time (RT) was generally less than one second for children in grades 4 to 6. The ECTs, administered to over 800 black and white elementary school children, strongly bore out Spearman's hypothesis, with a correlation of .80 between the extent to which these information processing measures measured g and the magnitudes of the black-white differences on these measures.[17]

Jensen's theory that the black-white difference in intelligence is principally a difference in Spearman's g has led him to modify his previous theory of differences in terms of Level I and Level II abilities advanced in the 1960s. His more recent formulation incorporates the earlier theory in so far as Level II abilities are better measures of Spearman's *g* than Level I abilities. This explains why the black-white difference is greater on Level II than on Level I abilities. Although Jensen considers that a difference in Spearman's *g* is the principal factor responsible for the black-white intelligence difference, he has found that this is not the sole factor. He has found that in addition to the *g* difference, blacks tend to have lower average scores on spatial visualization ability and higher average scores on short term memory.[18]

Jensen also showed that two processes known to be biological, regression to the mean and inbreeding depression, can be analyzed to demonstrate the genetic component of race differences in intelligence.

In 1994 Jensen retired from his professorship at the University of California, Berkeley, but he has continued his research on the nature of *g* and its educationally and socially significant correlates. In 1998 he published *The g Factor*, a book which places his own research on mental abilities within the context of historic and contemporary research on the *g* construct and its relevance to the human condition.[19] Over a period of more than 40 years Jensen has produced over 400 publications on the nature of intelligence and on the black-white difference. Jensen has made a major contribution to the advancement of knowledge on these problems, and his work has served as an inspiration to those who have followed in his footsteps.

Notes

1. Jensen, A. R. 1968. Patterns of mental ability and socioeconomic status. *Proceedings of the National Academy of Sciences.* 60: 1330-1337.

2. Jensen, A. R. 1973. Level I and level II abilities in three ethnic groups. *American Educational Research Journal.* 4: 263-276.

3. Jensen, A. R. 1969. How much can we boost IQ and scholastic achievement? *Harvard Educational Review.* 39: 1-123.

4. *Ibid.* 82.

5. Jensen, A. R. 1972. *Genetics and Education.* London: Methuen.

6. Jensen, A. R. 1973. *Educability and Group Differences.* London: Methuen.

7. *Ibid.* 363.

8. Jensen, A. R. 1980. *Bias in Mental Testing.* London, U.K.: Methuen.

9. *Ibid.* 740.

10. Jensen, A. R. 1981. *Straight Talk about Mental Tests.* New York, NY: Free Press.

11. Spearman, C. 1904. General intelligence objectively determined and measured. *American Journal of Psychology.* 15: 201-293.

12. Jensen, A. R. 1987. Individual differences in the Hick paradigm. In P. A. Vernon (ed.) *Speed of Information Processing and Intelligence.* Norwood, NJ: Ablex, 1987.

13. Reed, T. E. & Jensen, A. R. 1993. Choice reaction time and visual pathway nerve conduction velocity both correlate with intelligence but appear not to correlate with each other: implications for information processing. *Intelligence.* 17: 191-203.

14. Jensen, A. R. & Sinha, A. R. 1993. Physical Correlates of Intelligence. In P. A. Vernon (ed.) *Biological Approaches to the Study of Human Intelligence.* Norwood, NJ: Ablex.

15. Jensen, A. R. & Johnson, F. W. 1994. Race and sex differences in head size and IQ. *Intelligence.* 18: 309-334.

16. Jensen, A. R. 1985. The nature of the black-white difference on various psychometric tests: Spearman's hypothesis. *Behavioral and Brain Sciences*. 8: 193-219.

17. Jensen, A. R. 1993. Spearman's hypothesis tested with chronometric information processing tasks. *Intelligence*. 17: 47-78.

18. Jensen, A. R. & Reynolds, C. R. 1982. Race, social class and ability patterns on the WISC-R. *Personality and Individual Differences*. 3: 423-428.

19. Jensen, A. R. 1998. *The g-Factor: The Science of Mental Ability*. Westport, CT: Praeger.

Hans J. Eysenck, Ph.D.
Institute of Psychiatry,
University of London

Chapter 18

Hans J. Eysenck

Hans Eysenck (1916–1997) was one of the pioneers in the formulation of the taxonomy of personality and in the development of theory on its biological and genetic basis. He made numerous other important contributions to psychiatry over the course of more than half a century, including work on the neurophysiological processes underlying individual differences.

Hans Jurgen Eysenck was born in 1916 in Berlin. His parents were both actors. As an adolescent Eysenck's interests lay in science, and he planned to study physics at the University of Berlin. But when he was 17 years old Hitler came to power, and the Nazis put pressure on promising young men like Eysenck to join the party. Eysenck was strongly opposed to Nazi ideas and policies and decided to leave Germany. He went first to France and then to England. In 1935 he entered University College, London, to read psychology in the department headed by Sir Cyril Burt. He graduated in 1938 and obtained his Ph.D. in 1940. He worked initially as a research psychologist at the Mill Hill Hospital dealing with the psychological traumas suffered by servicemen during the war. In 1946 he moved to the Institute of Psychiatry in London as Director of the Psychology Department. He spent the remainder of his career at the Institute, at which he was appointed professor in 1955.

During his work of nearly 40 years at the Institute of Psychiatry, Eysenck made a number of important professional contributions to the development of psychology. He built up his department to become the largest postgraduate psychology center in London. He started a postgraduate course in clinical psychology which played a major role in establishing the profession; he introduced behavior therapy as an alternative to psychoanalysis, to which he has always been opposed, for the treatment of neuroses, and he founded two journals, *Behavior Research and Therapy*, and *Personality and Individual Differences*. Eysenck's principal interest was in writing and research. He wrote some 75 books and more than a thousand articles, which have had considerable influence in psychology.

EYSENCK'S PERSONALITY SYSTEM

In the 1940s through the 1960s Eysenck's principal research was concerned with personality. His first work dealt with the problem of how personality should be conceptualized. At this time there was no consensus on this issue. It was believed by some that personality could be analyzed in terms of different types such as the melancholic, phlegmatic, sanguine, and choleric personalities of ancient Greek medicine. This approach persisted in psychoanalytic theory, which posited such entities as the anal personality nurtured by over-rigorous toilet training. Others asserted that human personalities are unique and defy any attempt at categorization. Still others contended that the concept of personality was itself a snare and a delusion and that human behavior was determined by the immediate state in which the individual found himself, not any enduring traits.

Eysenck rejected these approaches in favor of personality trait theory. This holds that personality consists of a number of traits or, as he sometimes called them, "dimensions." Being dimensions, they are continuously distributed in the population, can be measured, and everyone can be given a score on them. These scores represent the individual's personality. In the 1940s and 1950s, Eysenck saw the main task of the development of a theory of personality as consisting of the identification and measurement of the major personality traits. In 1947 he identified two of these, which he called neuroticism and introversion-extraversion.[1] Neuroticism was defined partly in terms of psychoneuroses which are located at the high end of the dimension. Eysenck proposed that the independent trait of introversion-extraversion differentiated different kinds of psychoneurosis: those with anxiety and obsessional neurotics were conceptualized as introverted, hysterics as extraverted neurotics.

In 1952 Eysenck identified a third major personality trait which he called psychoticism.[2] This was shown to be independent of neuroticism, and introversion and was defined partly by psychotic mental illnesses at the high end of the dimension. Thus Eysenck proposed that both the neuroses and the psychoses are not qualitatively distinct illnesses but extreme forms of personality traits continuously distributed in the population.

Eysenck's three "superfactors" trait system has had considerable influence in personality theory although there was still, in the late 1990s, no agreed consensus among personality theorists on whether Eysenck was correct in limiting the number of personality traits to three (introversion-extraversion, neuroticism, and psychoticism). For several decades the principal rival personality theory was that of R. B. Cattell, whose system contains 23 traits, while a number of contemporary theorists favor what they call "the big five" personality theory which deletes Eysenck's psychoticism factor and adds factors called openness, agreeableness, and conscientiousness. Nevertheless, there is a wide consensus that Eysenck achieved a major advance in conceptualizing and measuring three of the most important traits which appear in virtually all contemporary personality theories. His personality questionnaires have been translated, standardized, and used for research and selection purposes in 37 countries.[3]

Eysenck has devoted some of his energies to exploring the associations between his personality traits and a variety of social behaviors and attitudes. For instance, he showed that extraverts tend to be tough-minded and authoritarian in their attitudes towards political and social issues.[4] Extraverts also tend to be more sexually permissive in their attitudes and behavior[5] and to be more likely to smoke than introverts.[6]

BIOLOGICAL AND GENETIC BASES OF PERSONALITY

Eysenck spent much of his career seeking the biological and genetic bases of his three personality traits.[7] He proposed that an individual's level of neuroticism is determined biologically by the sensitivity and reactivity of the sympathetic nervous system, which is responsible for anxiety reactions to stress. Eysenck and his colleagues and students carried out numerous experiments confirming this theory by showing that individuals' levels of neuroticism are related to their sensitivity to pain and neurophysiological reactivity to stimulation. Eysenck proposed that the biological basis of introversion-extraversion lies in the level of neurophysiological arousal of the brain stem reticular formation. Introverts are posited to have higher levels of arousal than extraverts. Eysenck tested and confirmed this theory in a number of studies showing that introverts have an advantage over extraverts in tasks requiring sustained attention and concentration. People's levels of neuroticism and introversion-extraversion are not, however, solely determined biologically. They are affected also by their life experiences and upbringing. Eysenck did not, however, formulate a theory of the neurophysiological basis for his trait of psychoticism.

Eysenck's theories concerning the biological bases of neuroticism and introversion-extraversion imply that these traits are to some degree determined genetically. Together with several colleagues, Eysenck carried out a number of studies to determine this. Two studies on twins carried out in the 1950s showed that both traits have high heritability. In a review of the literature published in 1985, written in collaboration with his son, Michael Eysenck, he concluded that the heritability of his three personality traits lies between 60 and 70 percent.[8]

THE PSYCHOLOGY AND HERITABILITY OF CRIME

Eysenck wrote extensively on the personality and genetic factors responsible for crime. His fullest treatment of this set of issues appeared in 1989 in a book written in collaboration with his colleague Gisli Gudjonsson.[9] They showed that the principal personality determinant of criminal behavior is a high level of the personality trait of psychoticism. This is envisaged as a continuum running from pro-social and well socialized behavior (low psychoticism) to anti-social and criminal behavior (high psychoticism). Many studies have found that males obtain higher average scores than females on psychoticism, and this is consistent with the higher rates of crimes committed by males. Crime is also associated with the personality traits of high neuroticism and extraversion, although these associations are less strong than the association between crime and psychoticism.

Eysenck argued that the criminal personality is determined by both genetic and environmental factors. Genetic factors exert major effects on an individual's level of psychoticism, and the level of psychoticism is a major determinant of crime. Hence genetic factors are likely to exert important effects on crime, acting through psychoticism. Eysenck reviewed evidence supporting the conclusion that crime is significantly affected by genetic influences. This evidence comes from studies comparing the similarity in respect of crime of identical twins compared with non-identicals, which has consistently shown greater similarity among identicals. In addition, there is evidence showing that adopted children resemble their natural parents in regard to crime more than they resemble their adoptive parents. Taking the evidence as a whole. Eysenck estimated the heritability of crime at 59 percent.

Eysenck believed that environmental factors also determine whether people become criminals. His theory was that parents differ in the effectiveness with which they condition their children to suppress anti-social behavior. This conditioning process leads to the development of the conscience. Parents who are ineffective at socializing fail to nurture their children's conscience, with the result that the children lack the moral restraints which control anti-social behavior in well-socialized individuals.

INTELLIGENCE AND RACE DIFFERENCES

Eysenck wrote on the subject of intelligence beginning in the 1950s. In 1971 he produced a short book *Race, Intelligence, and Education* for the general public in which he set out the evidence of the difference in average IQ between blacks and whites in the United States.[10] He broadly endorsed the conclusions of Audrey Shuey and Arthur Jensen to the effect that the weight of the evidence indicated that genetic factors make a significant contribution to this difference. He also argued that compensatory education had not yielded any lasting beneficial effects.

In 1979 Eysenck wrote *The Structure and Measurement of Intelligence*.[11] This was a textbook which set out the evidence on the hierarchical structure of intelligence consisting of Spearman's general factor, group factors, and narrower specifics; the evidence on the heritability of intelligence which concluded that the figure is approximately 80 percent; the nature of environmental effects; and the contribution of intelligence to socioeconomic status. This book also presented a brief discussion of eugenics in relation to intelligence in which the evidence for possible deterioration in genotypic intelligence was discussed. Some element of eugenic intervention was tentatively suggested, such as advice on better family planning:

There seems a case to be made for considering such eugenic measures both from a long term point of view and in the short term in order to reduce the social and economic cost of unwanted children. At the same time it makes good sense to encourage higher fertility at the high end of the IQ scale as well as to reduce it at the low end, perhaps by means of increased grants to married students or some other economic means.[12]

NEUROPHYSIOLOGICAL BASIS OF INTELLIGENCE

It was not until quite late in his career that Eysenck undertook research on intelligence. His particular interest was the hypothesis that the neurophysiological basis of intelligence lies in the speed of neural transmission through the nervous system. His work on this theory was supported by the Pioneer Fund from 1986.

The idea that the neurophysiological basis of intelligence might lie in the speed with which the brain is able to process information was originally advanced by Sir Francis Galton. Eysenck was one of the first to take up this idea. In 1967 he proposed a general theory of mental speed as the basis of intelligence.[13] The essence of the theory was that faster neural transmission would facilitate mental processes and produce better cognitive performance, which would be expressed in higher IQ scores.

From the early 1980s Eysenck and his colleagues carried out and published a number of studies supporting his theory. There have been three principal results of these studies. First, reaction times, normally measured by the speed of pressing a button when a light comes on, are positively correlated with intelligence. The theoretical explanation proposed for this is that reaction times provide a direct measure of the speed of information processing by the brain. Second, measures of more complex reaction times, for instance in the so-called odd-man-out task (where three lights appear, and the subject has to press a button beneath the one furthest from the remaining two), show higher correlations with intelligence than more simple reaction times. A probable explanation for this is that more complex reaction times require a longer time for neurological processing, in which speed differences between individuals will accumulate and produce higher correlations with intelligence. Third, the variability in reaction time responses shows a correlation with intelligence, possibly because faults in the speed of neural transmission sometimes produce exceptionally slow reaction times in less intelligent individuals and thereby greater variability. Eysenck summarized his work on reaction times and intelligence in a review paper published in 1993.[14]

Eysenck and his colleague Paul Barrett also worked on an even more direct test of the theory that speed of neural processing is the neurophysiological basis of intelligence. This consists of the measurement of the speed of neural transmission along the ulnar nerve.[15] The study found that speed of transmission had no relationship with intelligence, but there was a relationship with the variability of the transmission speed over a number of trials. This is consistent with the reaction time results, where variability has also been found to be associated with intelligence (the association is negative, that is to say variability is associated with lower IQ scores).

POLITICAL CORRECTNESS IN BRITAIN

Like several Pioneer-funded scholars in the United States, Eysenck was a victim of political correctness, the attempt to suppress the truth in the service of ideology. In 1973 he was physically attacked at the start of a lecture at the London School of Economics, and he was prevented from lecturing or shouted down at other universities. In 1990 a story critical of the Pioneer Fund appeared in a British newspaper, upon which the Dean of the Institute of Psychiatry prohibited Eysenck from receiving any further support from the Pioneer Fund through the Institute.

In spite of his willingness to put forth politically incorrect views when the evidence supported them, Eysenck received a number of awards and distinctions. He was the first president of the International Society for the Study of Individual Differences, received the Distinguished Scientist Award from the American Psychological Association, was nominated William James Fellow by the American Psychological Society, was given the presidential award for services to psychology by the American Psychological Society, and the first award by the International Society for the Study of Individual Differences for distinguished contributions to the study of personality, and received an award from the American Psychological Association for one of the greatest lifetime contributions to the field of clinical psychology. In 1990 a survey of American historians and chairpersons of the most influential psychologists in the world included Eysenck among the top ten.[16]

Hans J. Eysenck, one of the greatest figures in contemporary psychology, died on 4 September 1997 at the age of 81.

NOTES

1. Eysenck, H. J. 1947. *Dimensions of Personality*. London, U.K.: Routledge & Kegan Paul.
2. Eysenck, H. J. 1952. *The Scientific Study of Personality*. London, U.K.: Routledge & Kegan Paul.
3. Lynn, R. and Martin, T. 1995. National differences in thirty-seven nations in extraversion, neuroticism, psychoticism and economic, demographic and other correlates. *Personality and Individual Differences*. 19: 403-406.
4. Eysenck, H. J. 1954. *The Psychology of Politics*. London, U.K.: Routledge & Kegan Paul.
5. Eysenck, H. J. 1976. *Sex and Personality*. London, U.K.: Open Books.
6. Eysenck, H. J. 1980. *The Causes and Effects of Smoking*. London, U.K.: Temple Smith.
7. Eysenck, H. J. 1967. The Biological Basis of Personality. Springfield, IL: Charles C. Thomas.
8. Eysenck, H. J. & Eysenck, M. W. 1985. *Personality and Individual Differences*. New York, NY: Plenum.
9. Eysenck, H. J. & Gudjonsson, G. H. 1989. *The Causes and Cures of Criminality*. New York, NY: Plenum.
10. Eysenck, H. J. 1971. *Race, Intelligence and Education*. London: Temple Smith.
11. Eysenck, H. J. 1979. *The Structure and Measurement of Intelligence*. Berlin: Springer-Verlag.
12. *Ibid*. 174.
13. Eysenck, H. J. 1967. Intelligence assessment: a theoretical and experimental approach. *British Journal of Educational Psychology*. 37: 81-98.
14. Eysenck, H. J. 1993. The Biological Basis of Intelligence. In P.A. Vernon (ed.) *Biological Approaches to the Study of Human Intelligence*. Norwood, NJ: Ablex.
15. Barrett, P., Daum, I. & Eysenck, H. J. 1990. Sensory nerve conduction and intelligence. *Journal of Psychophysiology*. 1: 1-13.

16. Korn, J., Davis, R. & Davis, S. 1991. Historians and chairpersons' judgments of eminence among psychologists. *American Psychologist.* 46: 789-792.

P.A. Vernon, Ph.D.
University of Western Ontario

Chapter 19

P. A. Vernon

Philip Anthony (Tony) Vernon (born 1950) has worked on the neurophysiological basis of intelligence and reaction times, speed of neural conduction, and brain size. In addition, he has carried out genetic studies by testing twins and shown that these basic processes have a significant heritability.

Tony Vernon was born in 1950 in England. His father was the psychologist Philip E. Vernon, whose work is described in Chapter 16. Tony Vernon was educated up to the age of 18 in England. In 1968 he emigrated with his parents to Calgary, where he entered the university, from which he obtained his B.A. and M.Sc. He moved to the University of California to work for his Ph.D. under the supervision of Arthur Jensen. In 1980-1981 he taught in the School of Education at Berkeley. In 1982 he was appointed assistant professor of psychology at the University of

Western Ontario and was promoted to full professor in 1992. His work on neurophysiological processes underlying intelligence and on heritability has been supported by the Pioneer Fund since 1985.

INTELLIGENCE AND SPEED OF INFORMATION PROCESSING

From the 1960s onwards a number of psychologists have been interested in the possible relationship between intelligence and speed of information processing. The theory has been that intelligent people may process information more rapidly than less intelligent people and that this would be expressed in their having faster reaction times measured in simple tasks such as pressing a button when a light comes on. This theory would be substantiated by showing that there is a correlation between reaction times and intelligence. Vernon began to work on this problem in the 1980s. He showed that reaction times are significantly correlated with intelligence and that these correlations are greater for more complex reaction time tasks.[1] He has also shown that this association is present for both timed and untimed intelligence tests, showing that the common factor is not simply one of speed.[2] In a later study, tests of short term memory, together with tests of intelligence and reaction times, showed significant correlations between all three measures, supporting the concept of Spearman's general factor which expresses itself in the performance of all cognitive tasks.[3] However, statistical analyses suggested that general intelligence is not a unitary factor but comprises at least two components of short term memory and speed of information processing and that speed of information processing itself consists of the two uncorrelated components of general speed and long term memory.

Vernon has carried out a similar study involving the administration of tests of memory, reaction times, and intelligence to four to six year old children. In this it was found that reaction times were uncorrelated with intelligence, suggesting that speed of information processing is not relevant to the intellectual abilities of children at this age and only becomes relevant at some later stage of cognitive development.[4]

NERVE CONDUCTION VELOCITY AND INTELLIGENCE

Vernon has been one of several psychologists who have attempted to find an even more fundamental basis of intelligence in the properties of the nervous system. The general assumption has been that the brains of individuals with higher IQs are more efficient than those with lower IQs and that this must be due to some form of neurophysiological advantage. The problem has been to identify the specific neurophysiological advantage. One of the leading theories is that the neurophysiological basis of intelligence might lie in the speed with which nerve impulses are transmitted through the nervous system, including the brain. This is known as nerve conduction velocity. Individuals with fast nerve conduction velocity would have quicker reaction times and would be able to process information more rapidly, and this should be expressed in faster and more effective performance on intelligence tests. If this theory is correct, it would explain the well known association between fast reaction times and performance on intelligence tests, both of which would be expressions of nerve conduction velocity. Several psychologists have worked on this theory, including Pioneer grantees H. J. Eysenck, A. R. Jensen, T. E. Reed, as well as P. A. Vernon.

Vernon's first study was published in 1992 and showed positive results.[5] Nerve conduction velocity was measured in the arm on 85 students, and was found to be positively associated with intelligence at a correlation of 0.42. The students' reaction times were also correlated with nerve conduction velocity and with intelligence. This appeared to be a major breakthrough in the quest for the neurophysiological basis of intelligence. However, similar studies by Eysenck and his collaborators[6] and by Reed and Jensen[7] have failed to find this positive association. To check his result, Vernon carried out a further study on 38 young adult women, and on this occasion no association was obtained between nerve conduction velocity and intelligence.[8] Re-examination of the first study suggested that the positive correlation between nerve conduction velocity and intelligence is present in males but not in females and that this might account for the conflicting results in the different studies. If this is so, it would appear that the neurophysiological basis of intelligence differs for males and females.

BRAIN SIZE AND INTELLIGENCE

The question of whether brain size is related to intelligence was first considered by Sir Francis Galton in 1888, when he reported that Class A students at Cambridge had head sizes about 3 percent larger than Class C students.[9] Since that time a number of studies have been carried out on this issue and have typically reported positive correlations between head sizes and intelligence of around 0.2. Those who have worked on this issue have generally assumed that head size is an approximate measure of brain size, and hence that brain size is positively related to intelligence. The shortcoming of this work is that it is not clear how closely head size and brain size are related. If head size is only a rough approximation of brain size, the true correlation between brain size and intelligence might be substantially higher than the correlations between head size and intelligence.

A significant advance on this problem was made in 1991 by Lee Willerman and his colleagues in a study in which they obtained direct measures of brain size by the use of magnetic resonance imaging (MRI) and examined this in relation to intelligence in 40 students. They found a correlation of 0.35.[10] Two further studies using this technique obtained correlations of similar size. A fourth study using this method was published in 1994 by Vernon and his colleagues.[11] They reported results for 40 adult females for whom the correlation between brain size and IQ was 0.40. Subsequently, there has been an explosion of studies using MRI techniques, all finding an average of about 40.

Vernon's latest study, this time of 68 adult males, replicated the 0.40 correlation between brain volume and IQ. Like many other studies, it also found that head size variables show the expected positive correlation with IQ scores but at a lower magnitude (about 0.20). It was also found that brain size was correlated more highly with the g-component of IQ scores, and that the sizes of the left and right hemispheres were not associated with verbal or nonverbal abilities. The only substantive association appeared to be with total brain volume and general intelligence. In a further discussion of this issue, Vernon and his colleagues estimate the true correlation between brain volume and intelligence at 0.50.[12]

THE WESTERN ONTARIO TWIN STUDY

In the late 1980s Vernon set up the Western Ontario twin study. This consists partly of a longitudinal study of infant twins who are being tested and followed up over a period of years, and partly of a study of approximately 250 pairs of young adult twins and siblings. The objective of the infant twin study is to obtain estimates of the heritability of the motor, mental, and temperamental development of 150 pairs of infant twins and the degree to which this differs at different stages of infancy and childhood. The twins are first tested when they are three months old, at which time they are given the Bayley Scales of Infant Development; they are tested again at 6, 9, 12, 18, 24,

36, 48, 60, and 72 months of age. At the 6 and 9 month sessions, they are administered the Fagan attention-to-novelty test in addition to the Bayley tests. At the 24 month session they are given the Bayley and the Stanford Binet; at 36 months they are administered the Binet, and a variety of age-appropriate measures of their motor and temperamental development are taken; at the 48 month session they are administered the Wechsler Preschool Test and the Goodenough-Harris Drawing Test, writing samples are taken, and their parents complete a number of questionnaires about their personality; and at 60 and 72 months they are administered the Wechsler Intelligence Scale for children and the Raven Coloured Progressive Matrices, the Goodenough-Harris Drawing Test, and writing samples are taken. As to their mothers, when the twins are 18 months, the twins and their mother attend a test session at the university and other sessions are conducted in their own homes, where videotapes of mother-child and stranger-child interactions are made for subsequent analyses.

The major goal of this study is to attempt to identify factors that contribute to the differential personality and mental ability development of twins. By studying twins from very early in their lives, it may be possible to identify salient factors that operate at this stage and which affect their later development. Another goal of the study is the development of infant tests that serve as predictors of later intelligence. The Fagan test is one such infant predictor, and other information-processing tests have been included in the twins' three to nine month assessments.

The study of adult twins has involved the administration of tests of intelligence and of reaction times, to young adult MZ and DZ twins and non-twin siblings. In this study it has been found that reaction times have an appreciable heritability; 50 pairs of identical and 52 pairs of non-identical twins were tested for reaction times, and comparison of the twin correlations for various reaction times measured indicated an average heritability of 0.51.[13]

In the adult twins' study, the subjects complete a number of home and school environment and background questionnaires. One of the objectives of the project is to relate differences in twins' and siblings' environments to differences in their personalities. This is a cross-Canada study: data have been collected in Vancouver, Calgary, Toronto, London (Ontario) and Ottawa. Approximately half the sample of twins has also been administered a battery of creativity tests in order to investigate the genetic correlations between these measures and performance on the intelligence and personality tests. The twins study has been expanded into a collaborative project with Robert Derom of Catholic University, Robert Vlietinck of the University of Leuven in Belgium, and Evert Thiery of the University of Ghent. Derom has been affiliated with the East Flanders Prospective Twin Survey, a study with a sample of approximately 1,200 pairs of MZ twins, of whom half are 12 years old or older. A large amount of prenatal and postnatal data has been collected from these twins, including time and type of placentation.

In addition to his research papers, Vernon has edited three books on intelligence and personality: *Speed of Information Processing and Intelligence; Biological Approaches to the Study of Human Intelligence;* and *The Neuropsychology of Individual Differences.*[14]

NOTES

1. Vernon, P. A. 1986. The g-loading of intelligence tests and their relationship with reaction times: a comment on Ruchalla et al. *Intelligence*. 10: 93-100.

2. Vernon, P. A. & Kantor, L. 1986. Reaction time correlations with intelligence test scores obtained under either timed or untimed conditions. *Intelligence*. 10: 315-330.

3. Miller, L. T. & Vernon, P. A. 1992. The general factor in short term memory, intelligence and reaction time. *Intelligence*. 16: 5-29.

4. Miller, L. T. & Vernon, P. A. 1996. Intelligence, reaction time, and working memory in 4- to 6-year-old children. *Intelligence*. 22: 155-190.

5. Vernon, P. A. & Mori, M. 1992. Intelligence, reaction times and peripheral nerve conduction velocity. *Intelligence*. 16: 273-288.

6. Barrett, P. T., Daum, I. & Eysenck, H. J. 1990. Sensory nerve conduction and intelligence: a methodological study. *Journal of Psychophysiology*. 4: 1-13.

7. Reed, T. E. & Jensen, A. R. 1991. Arm nerve conduction velocity (NCV), brain NCV, reaction time and intelligence. *Intelligence*. 15: 33-47.

8. Wickett, J. C. & Vernon, P. A. 1994. Peripheral nerve conduction velocity, reaction time and intelligence: an attempt to replicate Vernon and Mori (1992). *Intelligence*. 18: 127-131.

9. Galton, F. 1888. Head growth in students at the University of Cambridge. *Nature*. 38: 14-15.

10. Willerman, L., Schultz, R., Rutledge, J. N. & Bigler, E. D. 1991. In vivo brain size and intelligence. *Intelligence*. 15: 223-228.

11.   Wickett, J. C., Vernon, P. A. & Lee, D. H. 1994. In vivo brain size, head perimeter and intelligence in a sample of healthy adult females. *Personality and Individual Differences*. 16: 831-838; Egan, V., Wickett, J. C. & Vernon, P. A. 1995. Brain size and intelligence: erratum, addendum and correction. *Personality and Individual Differences*. 19: 113-117.

12. Wickett, J. C., Vernon, P. A. & Lee, D. H., 2000. Relationships between factors of intelligence and brain volume. *Personality and Individual Differences.* 29. 1095-1122.

13. Vernon, P. A. 1989. The heritability of measures of speed of information processing. *Personality and Individual Differences.* 10: 573-576.

14. Vernon, P. A. (ed.) 1987. *Speed of Information Processing and Intelligence.* Norwood, NJ: Ablex; 1993. *Biological Approaches to the Study of Human Intelligence.* Norwood, NJ: Ablex; 1994. *The Neuropsychology of Individual Differences.* San Diego, CA: Academic Press.

T. Edward Reed, Ph.D.
University of Toronto

Chapter 20

T. Edward Reed

T. Edward Reed (born 1923) is a zoologist and geneticist who has made studies on the proportion of Caucasian ancestry in African Americans and on race differences in neurophysiological responses to alcohol. He has also worked on the speed of neural conduction along nerve fibers as a possible neurophysiological basis of human intelligence.

T. Edward Reed was born in 1923 and grew up in San Diego, California. He attended the California Institute of Technology and the University of California, Berkeley, from which he graduated B.A. in zoology. At Berkeley he was impressed by the geneticist Curt Stern and became interested in human genetics. He followed Stern's advice to go to the Galton Laboratory, University College, England, at that time a leading center for human genetics, to study for his Ph.D. under Lionel Penrose. He obtained his Ph.D. in 1952 and then joined the human genetics department at the University of Michigan, Ann Arbor, headed by James V. Neel. During his eight years in Ann

Arbor, among other projects, Reed studied the dominant genes for genetic diseases including Huntington's chorea and made estimates of their appearance by mutation.

In 1960 Reed moved to the University of Toronto to take up the position of associate professor in zoology and pediatrics. He became a full professor in zoology and anthropology in 1969 and remained at Toronto until his retirement in 1989. Between 1961 and 1974 he studied the population genetics of blood and serum groups, particularly for evidence of natural selection and as indicators of racial admixture. Some of these studies were carried out with grants from the U.S. National Institutes of Health. His most well known work published during these years was his calculation that the amount of white ancestry among the black population of the San Francisco Bay area was 22 percent.[1] He used the presence of the gene for the Duffy blood group to make this estimate, selecting this gene because it is virtually absent among pure African blacks, but relatively frequent among Caucasians. In 1992 he summarized and commented on other estimates of the proportion of white genes in the American blacks, which had reached a broadly similar conclusion.[2]

During the years 1973 to 1984 Reed worked on the genetic aspects of responses to alcohol in humans and mice. He reviewed racial differences in the rate of alcohol metabolism and concluded that the most striking racial differences in reactions to alcohol is the greater sensitivity to alcohol of East

Asians and South American Indians. He proposed that this sensitivity is caused by the high rate of aldehyde dehydrogenase-1 deficiency (ALDH) among East Asians and South American Indians, among whom it is present in about half the populations, although it is virtually absent among Caucasians.[3] ALDH deficiency causes flushing and dysphoria following the consumption of alcohol. Reed proposed that the high level of ALDH deficiency in East Asian and South American Indian populations is caused by a mutant gene. The deficiency makes the drinking of alcohol unpleasant, and Reed proposed that this explains the low average level of alcohol consumption among East Asians, although he noted that it has had little deterrent effect on consumption among many native American Indians.

These racial differences in metabolic reactions to alcohol are testimony to the reality of genetically determined race differences and belie the assertion frequently made that race is merely a social construct with no biological basis. Reed has also worked on the effects of alcohol on nerve conduction velocity in mice, supported by the Natural Sciences and Engineering Research Council of Canada.

Nerve Conduction Velocity and Intelligence

Reed's work on nerve conduction velocity in mice set him thinking about the possibility that nerve conduction velocity might be part of the neurophysiological basis of intelligence in humans. The idea was that fast neural conduction would be a component of a neurally efficient brain and that this would be expressed in superior intelligence test performance. In 1987-1988 he took a sabbatical year's leave and went to Berkeley to work on this idea with Arthur Jensen. Much of Reed's work on nerve conduction velocity and its relation to intelligence has been carried out in collaboration with Jensen and has been supported by the Pioneer Fund.

The methodology of Reed's work on nerve conduction velocity has been to administer electrical stimulation to the wrist and pick up the registration of this in the brain. The initial electrical stimulation is transmitted along the neural pathways until it reaches the brain, where it is recorded. The time taken for this transmission is the nerve conduction velocity. Reed has carried out similar experiments with visual stimuli. These provide measures of nerve conduction velocity through the brain, whereas the experiments involving stimulation of the wrist provide measures of nerve conduction velocity principally through the arm. In addition to these measures, Reed has obtained measures of his subjects' reaction times and intelligence, using a standard intelligence test.

In the first of these studies[4] carried out on 200 students it was found that nerve conduction velocity in the arm did not correlate with intelligence test performance. However, nerve conduction velocity in the brain did show a positive correlation with intelligence test results of 0.26 (adjusted to 0.37 when corrected for restriction of range) between nerve conduction velocity in the brain and intelligence on a sample of 147 students.[5] It was also found that mentally retarded individuals had exceptionally slow brain nerve conduction velocities. There was no correlation between reaction times and nerve conduction velocity in this study. This suggests that reaction times and nerve conduction velocity are two independent determinants of intelligence test performance.[6]

Further studies measuring nerve conduction velocity from the wrist to the cerebral cortex at stages along the neural pathway confirmed that the speed of transmission up the arm and into the thalamus in the midbrain were uncorrelated with intelligence, but a positive correlation with intelligence was present with speed of transmission from the thalamus to the cerebral cortex.[7] Thus the conclusion of this set of experiments was that it is only nerve conduction velocity in the brain that is associated with intelligence, not nerve conduction velocity in other parts of the nervous system. In a dispute with fellow Pioneer grantee J. P. Rushton over whether brain size is an important determinant of intelligence and of race differences in modern humans, Reed maintains that size is of relatively little significance and that the speed and accuracy of neural transmission is probably the more important neurophysiological determinant of intelligence.[8]

ENVIRONMENTAL EFFECTS OF NERVE CONDUCTION VELOCITY

Intelligence is determined both by genetic and environmental factors, and many hypotheses have been advanced as to what the relevant environmental factors are. One of these which has frequently been proposed is cognitive stimulation. Reed has investigated this possibility in studies of the effect of stimulation on nerve conduction velocity in mice. In his first study, he found that the nerve conduction velocity was greater among mice who had more cage mates.[9] He interpreted this as the effect of the stimulation provided by interaction with other mice. In a further study, Reed reared 29 mice for approximately six weeks in a stimulating environment containing a running wheel, ladder, and some empty juice cans, and 25 mice were reared in small cages without toys. Tests on the mice carried out after these two different experiences showed that those reared in the stimulating environment had greater nerve conduction velocity than those in the non-stimulating environment.[10] The results may be applicable to humans and provide support for the theory that babies and children provided with a stimulating environment may show an increase in intelligence.

HERITABILITY OF INTELLIGENCE

The classical methods for estimating the heritability of intelligence are studies of the similarity of identical twins reared in different families; the differences between identical and non-identical twins reared in the same family; and studies of adopted children showing that these tend to resemble their biological parents. All of these techniques have shown that intelligence has a moderate to high heritability, especially among adults. The validity of these techniques is accepted by Reed, but he has added two additional considerations which also point to a significant heritability for intelligence. He calls these the evolutionary argument and the neurophysiological argument.[11] The evolutionary argument is that modern humans developed greater intelligence than the ape-like ancestors from whom they have evolved over the last three million years or so. This must have taken place because those individuals with greater intelligence left more children than those with lesser intelligence. This process would only result in an increase in the intelligence of the species as a whole if intelligence has some degree of genetic determination. The evolution of intelligence from ape to man may have come to an end during the last 10,000 years or so, but intelligence could not have ceased to have some degree of heritability during this comparatively short period. Thus, the fact that intelligence has evolved shows that it must have some heritability.

An additional component of this argument is that the brain size of many species has increased during the course of evolution from the small brains of fish and reptiles to the large brains of mammals, particularly those of monkeys, apes, and humans. This development of brain size must have taken place because larger brains conferred greater intelligence and could only have occurred if the greater intelligence provided by large brains had some heritability.

Reed's neurophysiological argument for the heritability of intelligence is that intelligence requires the processing of information in the brain. This processing involves the movement of electrical potentials along and across neurons, and this occurs at varying speeds which can be measured by an individual's reaction time, consisting of pressing a button on the presentation of a light or sound. Reed notes that the speed of processing information in reaction time tasks is in fact correlated with intelligence measured by psychometric tests; and that the speed of information transmission through neural pathways depends on a variety of neurophysiological mechanisms such as axon diameter, the amount of myelin around the axon and the concentration of the proteins involved in neurotransmission. He argues that there is so much genetic diversity in humans that this is bound to include the neurophysiological processes involved in the speed of neuronal transmission.

There is further direct evidence that three measures of the speed of neuronal transmission have heritabilities. These are peripheral nerve conduction velocity, visual evoked potential latency, and reaction time. Since the neurophysiological processes underlying intelligence have some heritability, Reed concludes that intelligence itself must have some heritability. Reed has demonstrated that nerve impulse velocity has a genetic basis in mice.[12] His method consists of breeding different strains of mice and demonstrating that these inbred strains differ among themselves in their average nerve impulse conduction rates. His conclusion is that the heritability of nerve impulse conduction velocity in mice is approximately 0.4.

NOTES

1. Reed, T. E. 1969. Caucasian genes in American Negroes. *Science.* 165: 762-768.

2. Reed, T. E. 1992. Issues in estimating Caucasian admixture in American blacks. *American Journal of Human Genetics.* 51: 679.

3. Reed, T. E. 1985. Ethnic differences in alcohol use, abuse, and sensitivity: a review with genetic interpretation. *Social Biology.* 32: 195-209.

4. Reed, T. E. & Jensen, A. R. 1991. Arm nerve conduction velocity (NCV), brain NCV, reaction time and intelligence. *Intelligence.* 15: 33-47.

5. Reed, T. E. & Jensen, A. R. 1992. Conduction velocity in a brain nerve pathway of normal adults correlates with intelligence level. *Intelligence.* 16: 257-272.

6. Reed, T. E. & Jensen, A. R. 1993. Choice reaction time and visual pathway nerve conduction velocity both correlate with intelligence but appear not to correlate with each other: implications for information processing. *Intelligence.* 17: 191-203.

7. Reed, T. E. & Jensen, A. R. 1993. A somatosensory latency between the thalamus and cortex also correlates with level of intelligence. *Intelligence.* 17: 443-450.

8. Reed, T. E. & Jensen, A. R. 1993. Cranial capacity: new Caucasian data and comments on Rushton's claimed Mongoloid-Caucasian brain difference. *Intelligence.* 17: 423-424.

9. Reed, T. E. 1988. Narrow-sense heritability estimates for nerve conduction velocity and residual latency in mice. *Behavior Genetics.* 18: 595-603.

10. Reed, T. E. 1993. Effect of enriched (complex) environment on nerve conduction velocity: new data and review of implications for the speed of information processing. *Intelligence.* 17: 533-540.

11. Reed, T. E. 1990. Evolutionary and neurophysiological arguments for the heritability of intelligence. *Cahiers de Psychologie Cognitive.* 10: 659-667.

12. Reed, T. E. 1983. Nerve conduction velocity in mice: a new method with results and analysis of variation. *Behavior Genetics*. 13: 257-265.

Lloyd G. Humphreys, Ph.D.
University of Illinois at Urbana-Champagne

Chapter 21

Lloyd G. Humphreys

Lloyd G. Humphreys (born 1913) has made a number of empirical and theoretical contributions to the concept of intelligence. In the later part of his career, he developed particular interests in mathematical giftedness. He has also written on race differences in intelligence and educational attainment.

In the spring of 1936, Humphreys accepted an offer from Stanford University of admission to the Ph.D. program, with financial aid. Early the following fall, Jack Hilgard asked him to participate in his conditioning research. This led to joint publications and to a dissertation, for which he obtained his Ph.D. In 1938 he received a National Research Council Fellowship to work in experimental psychology with Clark Hull at Yale. Even though he had embarked on a career as an experimental psychologist, Humphrey's interest in individual differences continued. He published two papers in this area while at Stanford.

Humphreys spent his year at Yale doing experimental work on Hull's general theory. He left Yale in 1939 to accept an instructorship at Northwestern University. In 1941, he obtained leave to take up a postdoctoral Carnegie Fellowship in the anthropology department at Columbia University. In April 1942 he left Columbia to take a commission in the U.S. Army Air Force, with the unit of the Aviation Psychology Program headed by J. P. Guilford. For the next three and a half years he gained extensive practical experience in research on human abilities.

After the war, Humphreys returned briefly to Northwestern as an assistant professor of psychology, before taking up an appointment as associate professor and director of testing and guidance at the University of Washington. In 1948 he became an associate professor of education and psychology at Stanford. Between 1951 and 1957 he served as research director in the personnel laboratory of the Air Force Personnel and Training Research Center at Lackland Air Force Base. In 1957 he was appointed to a full professorship at the University of Illinois, becoming head of the psychology department two years later.

By this time Humphreys was fully committed to research in individual differences. He has participated in the measurement program at Illinois continuously since 1957. He has stronger applied interests than most of his measurement colleagues. After his retirement he has continued his research, particularly on the identification and development of gifted children. His work has been supported by the Pioneer Fund since 1990.

THE NATURE OF INTELLIGENCE

Humphrey's views on the concept of intelligence have been broadly in the tradition of the British School of Charles Spearman, Sir Cyril Burt, Philip Vernon, and Raymond Cattell. In the late 1960s he published an analysis of data confirming the existence of Spearman's general intelligence factor (g) and its division into two subfactors of verbal-educational abilities and reasoning-spatial abilities.[1]

In subsequent papers he has analyzed further data sets demonstrating the presence of Spearman's general factor and a number of smaller factors.[2] He believes the evidence shows that intelligence is an important determinant of performance in educational and occupational attainment. He also accepts that individual differences in intelligence are to some extent determined by genetic factors but he is uncomfortable about attempts to be too precise about the magnitude of the genetic contribution which he places at somewhere between 20 percent and 80 percent. He has defined intelligence as:

> the acquired repertoire of all intellectual (cognitive) skills and knowledge available to the person at a particular point in time.[3]

Humphreys is concerned with measured (phenotypic) intelligence rather than with hypothetical genetic factors.

MATHEMATICAL GIFTEDNESS

One of Humphreys' main interests has been in the mathematically gifted. He has investigated the physical health of the mathematically gifted and also the disproportionate numbers of boys among the mathematically gifted. So far as physical health is concerned, he has examined the top 1 percent for mathematical ability among approximately 100,000 tenth graders and found that the mathematically gifted boys and girls had a higher incidence of myopia (short sightedness). Apart from this they tended to be more healthy than average. They had had fewer visits to the doctor, stomach troubles, rheumatic fever, or severe headaches. There were no differences between the mathematically gifted and other adolescents in respect of speech or hearing difficulties, mumps, asthma, hay fever, heart trouble, or colds.[4]

Humphreys has also been interested in the problem of why boys in middle to late adolescence tend to perform better in mathematics than girls. In collaboration with David Lubinski, he examined a sample of approximately 100,000 tenth graders given a number of tests of ability and educational attainment in 1960.[5] They found that boys obtained a slightly higher mean (.18 standard deviations) than girls and also that the range of mathematical ability was greater among boys. These two male advantages in combination produce a large excess of males at high levels of ability of the order of 10 males to one female.

ETHNIC DIFFERENCES IN "INADEQUATE LEARNING SYNDROME"

In 1988 Humphreys wrote a general review of the incidence of what he called "Inadequate Learning Syndrome" (ILS) among the major ethnic and racial groups in the United States.[6] The Inadequate Learning Syndrome consists of deficits in educational attainment in reading, arithmetic, mathematics, science, and other school subjects. His conclusion was that Asians and whites score at about the same level, Mexican Americans and native American Indians score lower, and blacks and Puerto Ricans obtain the lowest average scores. There is some evidence from the National Assessment of Educational Progress studies that blacks and Hispanics improved in reading relative to whites over the period 1970 to 1984. Similar small gains are evident in the Graduate Record Examination.

Humphreys is agnostic as to whether the differing incidence of Inadequate Learning Syndrome in different ethnic and racial groups is caused by environmental or genetic factors, or by some mix of the two. Even if the causes are wholly environmental, he is pessimistic about the likelihood of redressing them by interventions such as greater financial provision for schools, affirmative action, and so on. He regards the deficits as caused by deep-rooted cultural problems of welfare dependence, crime, teenage pregnancy, drugs, and other features of the underclass subculture which are more difficult to deal with by governmental action.

AFFIRMATIVE ACTION

Humphreys has opposed affirmative action policies designed to favor blacks with lower test scores than whites in college admissions and hiring.[7] He believes that there are real average differences in phenotypic (measured) intelligence between blacks and whites in the United States and that this is the principal factor responsible for their different average levels of performance in educational and occupational attainment. He is agnostic on the issue of whether the black-white differences are genetic or environmental or some mix of both, writing that:

> There are no dependable data concerning the size of a possible genetic contribution to racial differences in cognitive abilities.[8]

In spite of his open-mindedness on the issue of causation, Humphreys has produced data which throw doubt on a solely environmental determination of the black-white difference in intelligence. Many equalitarians have maintained that blacks perform poorly on intelligence tests because they are of lower average socioeconomic status than whites. But in 1972 Humphreys showed that the pattern of intellectual abilities differentiating the social classes is not the same as the pattern differentiating blacks and whites.[9] He demonstrated that among whites the higher socioeconomic classes are strongest on verbal abilities and less strong on the visuospatial abilities, while conversely the lower socioeconomic classes are weakest on verbal abilities and relatively stronger on the visuospatial abilities. If blacks were like lower socioeconomic status whites, they too would be weakest on the verbal abilities and stronger on the visuospatial abilities. But Humphreys' research showed this is not the case, and that blacks are, if anything, stronger on the verbal than on the visuospatial abilities. The conclusion that has to be drawn is that blacks are not like lower socioeconomic status whites.

Whatever the causes of the black-white differences in intelligence, Humphreys believes that they are not easily eliminated. He argues that affirmative action to favor blacks over better qualified whites will do little good and much harm, because it is likely to lead to an increase in racial tension. Affirmative action also fosters the belief among blacks that their poorer performance is caused by whites, and this is not the case.[10]

A COMMON DESTINY

Humphreys is critical of mainstream American social science for its refusal to recognize the contribution of intelligence and personality traits to educational and occupational achievement and its preference for explaining individual and race differences in achievement solely in terms of sociological concepts like poverty and racism. A typical mainstream social science analysis of the kind to which Humphreys objects is *A Common Destiny*.[11] This is a report published in 1989 of a committee of 22 social scientists set up by the National Research Council to consider the under-performance of blacks in the United States and to propose solutions. Humphreys has written a critique of the Report in which he maintains that the committee responsible for it ignores the psychological evidence of the contribution of intelligence to individual and racial differences in achievement.[12]

The Report concludes that improvement in the social and economic position of blacks in the United States has been disappointingly slow over the period of 35 years following the Supreme Court's decision of 1954 mandating the desegregation of schools in the case of *Brown v. Board of Education*; that inequalities between blacks and whites will continue in the future; and that the high prevalence of drug abuse, crime, teenage parenthood, poor educational attainment, and high rates of unemployment are likely to remain among inner city black communities. The Report made a number of recommendations for policies to improve the social and economic position of blacks, including provision of better education and health care, more employment opportunities, the reduction of discrimination and racial segregation in employment, education, and housing, and the improvement of social welfare programs.

Humphreys criticizes the Report for ignoring the evidence on the differences in intelligence between blacks and whites and the contribution of these differences to their relative social and economic performance. He notes that the word "intelligence" never appears in the body of the report and even the black-white difference on achievement tests receives little coverage. When the committee did consider achievement tests it entered a caveat about the tests being biased, but Humphreys believes that the research evidence shows that bias introduces only trivial distortion effects on test scores.

Furthermore, the Report nowhere considers the possible contribution of genetic factors to the failure of blacks to achieve social and economic equality with whites. Humphreys says that this is typical of social scientists who tend to be

> strongly environmentalist in their orientation towards causation... and naively environmental in their optimistic expectations about obtaining significant change through environmental manipulation.[13]

He does not claim that it has been proven that genetic factors are involved but he does suggest that

> no scientist qua scientist can reject the hypothesis that there is a genetic contribution to the race difference,[14]

and that the committee should have recognized and discussed this possibility. The Report blames poverty as the root cause of the social and economic under-performance of blacks without giving any adequate analysis of why blacks are disproportionately poor.

Humphreys argues that low intelligence is a significant cause of black poverty and that this has to be considered in any sophisticated account of the causes of disproportionate poverty rate among blacks. He singles out the Report's chapter on education for what he calls its "naive environmentalism." The Report argues that much of the under-performance of blacks is due to the poor quality of their education and that this could be improved by giving their schools greater resources, but Humphreys argues that the research evidence shows that education as such makes little contribution to race differences in occupational achievement.

Humphreys raises the question of why so many social scientists refuse to recognize the importance of intelligence as a determinant of educational and occupational achievement. He suggests two answers. First, once the role of intelligence is recognized it becomes necessary to confront the black-white IQ difference as a major contributor to the race difference in educational and occupational attainment. Second, recognizing the importance of intelligence can have the appearance of "blaming the victim" especially given the evidence for a genetic contribution to the differences. Humphreys concludes that the refusal of many social scientists to admit the importance of intelligence as an important determinant of social success and failure is politically and ideologically motivated.

Turning to the Report's policy recommendations, Humphreys argues that there are more problems in achieving the ideal of racially integrated schools than the Report's authors recognize. He says that because the average educational achievement of black children is lower than that of whites, the influx of large numbers of black children into all white schools will inevitably reduce the educational standards in the schools. White parents understand this and frequently move from racially-mixed cities to largely white suburbs to avoid integrated schools. This can be prevented by busing, but this is an interference with civil liberties. Even when implemented, these programs have not produced the amount of integration hoped for.

Humphreys is also critical of the Report's recommendation for affirmative action. He argues that an organization that appoints less qualified individuals inevitably reduces its efficiency and productivity, and universities practicing affirmative action jeopardize their academic standards. Because relatively few blacks score at the highest level, the only way top universities can admit large numbers of blacks is by creating a separate and lower standard of admission for them. This creates several problems. It causes resentment among whites who are refused admission; the blacks who are admitted under such programs perform less well than individuals admitted on the basis of merit and often demand concessions, such as being allowed to retake failed courses and the introduction of less academically demanding degree programs. For instance, after Harvard Medical School adopted an affirmative action policy of admitting under-qualified blacks in 1968 it has had to allow failing students to take more repeat examinations and has ceased to publish the students' distribution of scores on the National Board Examinations because it does not wish it to be known that these have fallen.

Humphreys also questions the Report's recommendation for more racially integrated housing. He argues that the preference of whites for living in all-white neighborhoods is not so much racist as based on the knowledge that racially integrated neighborhoods are less safe. He notes that upper-middle-class white liberals rarely support the introduction of integrated housing into their own neighborhoods. He suggests that the prototypic white liberal is Senator Edward Kennedy who advocates racially integrated schools for South Boston but sent his own children to white private schools. Humphreys concludes that the Report is too ready to blame "white racism" for the underachievement of blacks and fails to examine the psychological and biological factors responsible for individual and race differences in achievement in education and socioeconomic status.

In addition to his research, Humphreys has made a number of professional contributions to psychology. He served as editor of the Psychological Bulletin between 1964 and 1969, and of the *American Journal of Psychology* between 1968 and 1979. He was a member of the board of directors of the American Institutes for Research; a member of the Commission on Human Resources of the National Academy of Sciences, National Research Council (1972-1982); and chairman of Division J of the American Association for the Advancement of Science (1980). He was a member of the board of directors of the American Psychological Association (1975-1978) and of the Board of Human Resources Data and Analyses, National Research Council, National Academy of Sciences (1974-1977). He also has been assistant director for education of the National Science Foundation; vice president and chairman of Division I of the American Association for the Advancement of Science; and chairman of the Conference of Graduate Training Departments of Psychology (1962-1966).

Humphreys has served as president of the Division of Military Psychology of the American Psychological Association, the Division of Evaluation and Measurement of the American Psychological Association, and the Psychometric Society. He helped found and later served as chairman of the governing body of the Psychonomic Society. In 1995 he was given the annual award of the Educational Testing Service for achievement in psychometrics.

NOTES

1. Humphreys, L. G. 1967. Critique of Cattell's "Theory of fluid and crystallized intelligence: A critical experiment." *Journal of Educational Psychology.* 58: 120-136.

2. Humphreys, L. G. 1979. The construct of general intelligence. *Intelligence.* 3: 105-120.

3. Humphreys, L. G. 1994. Intelligence from the standpoint of a pragmatic behaviorist. *Psychological Inquiry.* 5: 179-192. p. 180.

4. Lubinski, D. & Humphreys, L. G. 1992. Some bodily and medical correlates of mathematical giftedness and commensurate levels of socioeconomic status. *Intelligence.* 16: 99-115.

5. Lubinski, D. & Humphreys, L. G. 1990. A broadly based analysis of mathematical giftedness. *Intelligence.* 14: 327-355.

6. Humphreys, L. G. 1988. Trends and levels of academic achievement in blacks and other minorities. *Intelligence*: 12: 231-260.

7. Humphreys, L. G. 1993. Predictively, ability tests are color blind. *Academe.* May: 35-36.

8. *Ibid.* 35.

9. Humphreys, L. G., Fleishman, A. I. & Lin, P. C. 1977. Causes of racial and socioeconomic differences in cognitive tests. *Journal of Research in Personality.* 11: 191-208.

10. Humphreys, L. G. 1994. Intelligence from the standpoint of a pragmatic behaviorist. *Psychological Inquiry.* 5: 179-192.

11. Jaynes, G. D. & Williams, R. M. 1989. (eds.) *A Common Destiny: Blacks and American Society.* Washington, D.C.: National Academy Press.

12. Humphreys, L. G. 1991. Limited vision in the social sciences. *American Journal of Psychology.* 104: 333-353.

13. *Ibid.* 335.

14. *Ibid.* 334.

Douglas N. Jackson, Ph.D.
University of Western Ontario

Chapter 22

Douglas N. Jackson

Douglas N. Jackson (born 1929) is the foremost authority in Canada on the construction of psychological tests of personality and intelligence. He has constructed the Jackson Personality Inventory, a questionnaire measure of personality traits.[1] He has also published more than 175 papers in professional journals, many of them on technical aspects of test construction, and also on personality disorders, vocational interests and job performance and satisfaction, and sex differences in cognitive abilities.

Douglas N. Jackson was born in 1929 and took his first degree in psychology at Cornell University in 1951. After graduating, he moved to Purdue University, where he obtained his M.Sc. in 1952 and Ph.D. in 1955. He then took up a post-doctorate fellowship in clinical research at the Menninger Foundation. During 1962-1964 he worked at Stanford University as a research fellow and associate professor. In 1964 he was appointed senior professor of psychology at the University of Western Ontario, where he has spent the remainder of his career.

SCIENTIFIC CREATIVITY

One of Jackson's primary interests has been in scientific creativity. In the book on this subject he edited with J. P. Rushton, Jackson contributed a chapter in which he discussed the values, personality traits, and motives of creative scientists and the type of environment most conducive to their productivity.[2] He argued that strong theoretical interests, curiosity, and a need for independence are salient characteristics of creative scientists. He also proposed that scientists score high on achievement motivation and that this concept is a product of two independent traits: the need for excellence, and acquisitiveness. Jackson argues that organizations need to be more aware of how to provide the best conditions for creative scientists to work in. These are the recognition of inventive productivity, freedom from bureaucratic interference, good financial rewards, and incentives for career advancements.

SEX DIFFERENCES IN INTELLIGENCE

Jackson received a grant from the Pioneer Fund in 1991 to study sex differences in intelligence and in personality. His approach was that males and females probably differ in the strength of different cognitive abilities and in personality traits, and the objective of the research was to obtain and analyze data to throw further light on this issue, to consider how far the sex differences have a genetic basis, what evolutionary selection pressures may have been responsible for them, and whether they may have evolved differently in different races as a result of differential selection pressures.

Jackson's first report on these questions was published in 1994 and consisted of an analysis, carried out in collaboration with Heinrich Stumpf, of the test scores of approximately 97,000 male and 90,000 female applicants for medical school in Germany over the years 1981 through 1989.[3] The applicants took 15 tests covering a wide range of cognitive abilities. These were reduced to three factors identified as reasoning ability, perceptual speed, and medium term memory. The results showed that males outperformed females on reasoning ability by .56 of a standard deviation, equivalent to 8.4 IQ points, while females outperformed males on memory by half a standard deviation, equivalent to 7.5 IQ points. There was no sex difference on the perceptual speed factor. The male advantage on reasoning and the female advantage on memory replicate other studies carried out in the United States, Britain and elsewhere, although the absence of any sex differences on perceptual speed is unusual because females generally perform better on this ability.[4] The reason for this result is that the German tests of perceptual speed contained spatial items on which males invariably perform better, which counterbalanced the female advantage in non-spatial perceptual speed tests.

The size of the male advantage on reasoning in this data set is rather larger than those found in studies in other countries, which have typically found a male advantage of the order of 3 to 4 IQ points.[5] Possibly the nature of the sample might explain this large difference, because applicants to medical school are not a representative sample of the population.

One of the most striking results of the study is the large female advantage in medium term memory. Little previous research has been published on this. Several studies on short term memory involving immediate memory for digits have found small female advantages of around 2 IQ points, only about a quarter of the size of the difference found in Jackson's study.[6] Probably different mental processes are involved in short term and medium term memory, and the female advantage is greater for the medium term ability. Also measured in the study were male-female differences in the variability of the test scores. It has frequently been proposed that males show greater variability than females. However, no general trend of this kind was present in the German data.

Finally, Jackson and Stumpf[7] examined the data to determine whether sex differences were decreasing over the time period. They found that there was no evidence that this was the case, contrary to the claims that have sometimes been made.[8] At the time of this writing Jackson has not completed his work on the evolution of sex and age differences in intelligence.

Douglas Jackson has received a number of honors. He is a Fellow of the Royal Society of Canada. He has served as president of the Division of Measurement, Evaluation and Statistics of the American Psychological Association (1989-1990) and president of the Society of Multivariate Experimental Psychology (1975-1976). He has worked as visiting professor of psychology at Pennsylvania State University (1978-1979) and at the University of Iowa (1983).

NOTES

1. Jackson, D. N. 1976. *The Jackson Personality Inventory*. Port Huron, MI: Research Psychologists Press; Jackson, D. N. 1984. *The Multidimensional Aptitude Battery*. Port Huron, MI: Research Psychologists Press.

2. Jackson, D. N. 1987. Scientific and technological innovation: its personological and motivational context. In D. N. Jackson & J. P. Rushton. 1987. (eds.) *Scientific Excellence: Origins and Assessment*. New York, NY: Sage.

3. Stumpf, H. & Jackson, D. N. 1994. Gender-related differences in cognitive abilities: evidence from a medical school admissions program. *Personality and Individual Differences*. 17: 335-344.

4. Born, M. P., Bleichrodt, N. & van der Flier, H. 1987. Cross-cultural comparison of sex-related differences on intelligence tests. *Journal of Cross-Cultural Psychology*. 18: 283-314.

5. *Ibid*.

6. *Ibid*.

7. Stumpf, H. & Jackson, D. N. 1994. *Op. cit*.

8. Feingold, A. 1988. Cognitive gender differences are disappearing. *American Psychologist*. 43: 95-103.

PART IV:
Behavioral and
Medical Genetics

Thomas J. Bouchard, Jr., Ph.D.
University of Minnesota

Chapter 23

Thomas J. Bouchard, Jr.

Thomas J. Bouchard, Jr. (born 1937) has been one of the leading researchers to demonstrate that genetic influences have important effects on intelligence and a number of personality traits. His method has been to study the degree of similarity of identical and non-identical twins separated shortly after birth and reared in different families. The results of his studies have both consolidated and extended existing evidence showing that genetic factors are important in the determination of these attributes. His work has also shown that the relevant environmental influences are unique to the individual rather than operating as family effects arising from shared or common family influences such as parental styles of upbringing, discipline, role models, encouragement, and the like.

  Thomas J. Bouchard, Jr. was born in 1937 in Manchester, New Hampshire. On leaving high school he served in the United States Air Force for three years, 1955-58. He took his first degree in psychology at the University of California, Berkeley, in 1963, and his doctorate from the same institution in 1966. His first teaching position was as a graduate student at Berkeley in 1963, where he was engaged as a research assistant (1963-1964), and where he was awarded a regent's fellowship (1964-1965) and a predoctoral fellowship (1965-1966). During the 1960s Bouchard participated in the free speech protest movement. He was arrested during the takeover of Sproul Hall, the Berkeley campus administration building, and spent one night in the Santa Rita prison after which his wife, who was caring for their first child, borrowed money from a friend and bailed him out. He was eventually tried, found guilty, and put on probation, in company with a few hundred other students.

In 1966 Bouchard took up an appointment as assistant professor in psychology at the University of California, Santa Barbara. In 1969 he moved to the psychology department at the University of Minnesota. He was appointed associate professor in 1970, and full professor in 1973. He served as the chairman of the psychology department from 1985 to 1991. At Minnesota he also served as chairman of the graduate school research advisory committee between 1986 and 1991, and as a member of the executive committee of the Institute of Human Genetics from 1985 to the present. An editor of *Behavior Genetics* between 1982 and 1986, associate editor of the *Journal of Applied Psychology*, and book review editor of *Social Biology*, Bouchard has written over a hundred articles in journals and books.

During the 1970s Bouchard's principal academic interests were in the psychology of small groups, the effects of high noise levels on adjacent residences and the incidence of crime as assessed with victimization surveys. It was not until the end of the decade that he became actively interested in the genetics of intelligence and personality. His first study in this area appeared in 1977 and was concerned with the X-linked recessive gene hypothesis of spatial ability which had been advanced to explain the male superiority in spatial ability. This study, carried out in collaboration with his graduate student Mark McGee, failed to support the hypothesis.[1]

THE MINNESOTA STUDY OF IDENTICAL TWINS REARED APART

In 1979 Bouchard began his study of the intelligence and personality of twins reared apart, work which continues to be his primary interest. In the conduct of the study he has been assisted by David Lykken, Matthew McGue, Auke Tellegen, and Nancy Segal, as well as numerous others who have worked with the research team and collaborated in the analyses of various parts of the data. The basic methodology of the Minnesota Twin study has been to obtain pairs of identical and non-identical twins who were separated shortly after birth and brought up in different families. The twins have come from various parts of the world and been brought to Minnesota for approximately a week, where they have undergone about 50 hours of psychological and medical testing. The diagnosis of whether the twins were identical or nonidentical has been carried out by the analysis of blood groups, four serum proteins, six red blood cell enzymes, fingerprint ridge count, ponderal (height-to-weight) index, and cephalic index. By the mid-1990s 62 pairs of identical twins and 43 pairs of non-identical separated twins had been studied. The Pioneer Fund has supported Bouchard's twin study from shortly after its inception. The Fund has provided in excess of 1 million dollars, more than it has given to any other research program. Bouchard's work has also received financial support from the University of Minnesota Graduate School, the Spencer Foundation, the Koch Foundation, the Seaver Institute, and the National Science Foundation.

INTELLIGENCE OF IDENTICAL TWINS REARED APART

The logic of the study of the IQs of identical twins reared apart to estimate the genetic effects on intelligence is that these twins have the same heredity but different environments. The degree to which they are similar is therefore a measure of the magnitude of the genetic effect. The degree of similarity is measured by the correlation between the twin pairs.

There have been five studies which are generally accepted as valid of the intelligence of identical twins reared apart. The first of these was carried out by Newman, Freeman, and Holzinger in the United States in the 1930s and obtained a correlation of .71 between the twin pairs; the second was carried out by Shields in Britain in the late 1950s and obtained a correlation of .75; the third was conducted in Denmark by Juel Nielsen, for which the correlation was .69; the fourth was carried out by N. L. Pedersen and her colleagues in Sweden and obtained a correlation of .78; the fifth was carried out by Bouchard and his colleagues and consisted of twins from various parts of the world and produced a correlation of .75.[2] Bouchard has calculated the average of all these studies, weighting the result of each study by the numbers of twin pairs, and arrives at the same figure of .75. This figure can be interpreted as a percentage and indicates that genetic factors account for 75 percent of the variance in intelligence. However, intelligence tests are not wholly reliable measuring instruments because of errors of various kinds that creep into the scores obtained. The reliability, or stability as it is sometimes called, of the IQ score (if given to the same person at different times) is estimated by Bouchard at .90. The correlation between the twin pairs can be corrected for the test reliability. The correction raises the correlation from .75 to .83. This is the best estimate of the heritability of intelligence derived from the identical twin method. The total number of pairs in the five studies is 162, of which 48 come from Bouchard's sample, making it the largest of the studies.

In addition to these five studies, a further study by the British psychologist Sir Cyril Burt reported that 53 pairs of separated identical twins showed a correlation for IQ of .77.[3] Burt's study has been criticized because of apparent errors in some of his figures, and for this reason is generally omitted from compilations of the research literature. Notice, however, that Burt's result of a correlation of .77 is indistinguishable from the figure of .75 obtained from the other five studies.

The identical twins reared apart method of estimating the genetic contribution to variability in IQ has been criticized by various writers, notably Leon Kamin.[4] Bouchard has examined these criticisms, which he describes as "pseudoanalysis" rather than reasonable evaluations and provided detailed refutations of each of them.

OTHER ESTIMATES OF THE HERITABILITY OF INTELLIGENCE

Bouchard has shown that other estimates of the heritability of intelligence agree closely with the figure of around 80 percent derived from identical twins reared apart. In 1981, in collaboration with Matt McGue, he reviewed the research literature on a variety of familial correlations for IQ, such as those between parents and children, siblings, half-siblings, cousins, and so forth.[5] All of these can be used to estimate the heritability of intelligence. Two of these correlations are particularly informative. The first is a comparison between the correlations of identical and same sex non-identical twins reared in the same family. In his 1981 review, the average correlations obtained from all studies were .86 for identical twins and .60 for non-identical twins. The heritability is obtained by doubling the difference between the two correlations which comes to .52.

This figure is considerably lower than the uncorrected correlation of .75 obtained for identical twins reared apart. In a later paper, Bouchard and his colleagues have shown that the reason for this discrepancy lies in the ages of the twins. It appears that the difference between the correlations of identical and non-identical twins is relatively small in childhood and adolescence but becomes greater among adults.[6] Thus, among four-to-six-year olds, the two correlations are .78 and .58, indicating a heritability of 40 percent. But among adults the two correlations are .88 and .50, indicating a heritability of .76 percent. Notice that the figure for adults is virtually identical to the .75 correlation for separated identical twins. This analysis shows that the genetic effects on intelligence are weaker among young children than they are among adults. Conversely, it shows that the environmental effects on the intelligence of children are relatively strong, but weaken progressively throughout later childhood and adolescence and are weakest amongst adults.

Bouchard's work on the IQs of separated identical twins confirms the results of the four earlier studies and also that of Burt, and of the studies comparing identical twins and non-identical twins reared in the same families in showing that the heritability of intelligence estimated from correlations uncorrected for reliability is approximately 75 percent, and that when this correlation is corrected for reliability it rises to approximately 83 percent. This result has recently been replicated in Sweden by a study of 93 identical and 113 non-identical adult twins reared together, in which the heritability of intelligence is estimated at 81 percent.[7]

Corroborating evidence for the heritability of intelligence comes from the zero correlation found between pairs of unrelated children reared by the same adoptive parents. On the basis of 108 cases in the research literature, Bouchard calculates this correlation as -.02.

This effectively zero correlation confirms the conclusion from the twin studies that common environmental effects on intelligence must be low, because if they were appreciable a positive correlation would be present. Bouchard's interpretation of the zero correlation is that the environmental factors affecting intelligence are not the usually supposed family influences of how much parents read to their children, what toys they buy them, the extent to which they take them to museums, the schools they send them to, and so on. If these factors had any effect they would pull the IQs of the unrelated pairs into positive correlation. The fact that the correlation is zero shows that the environmental factors affecting intelligence must be unique to the individual or what geneticists call "non-shared environmental factors." They are probably such things as the adequacy of prenatal nutrition, birth injuries, and diseases.

INHERITANCE OF PERSONALITY

In addition to their work on the heritability of intelligence Bouchard and his colleagues have used their separated identical twins to study the genetic contribution to personality. In 1988 they reported the results for 44 identical and 27 non-identical twins reared apart, together with a further 217 identical and 114 non-identical twins reared together, tested with the Multidimensional Personality Questionnaire, a measure of 11 personality traits.[8] The estimated heritabilities of the seven traits are: well being (.48); social potency (.54); achievement (.39); social closeness (.40); stress reaction (.53); alienation (.45); aggression (.44); control (.44); harm avoidance (.55); traditionalism (.45); and absorption (.50). All the heritabilities fall within the range of .39 and .55, indicating that about half of the variance in all the traits is genetically determined. The analysis also divided the environmental contribution into the effects of shared family environment and unshared environment. Perhaps surprisingly, the shared family environment consisting of the way parents bring up their children, their style of discipline, the role models they provide, and so forth that the children reared in the same family share, made no significant contribution to the variance in any of the personality traits. The significant environmental factors were of the unshared type, i.e., those which impacted on one twin but not on the other. It is unclear what these unshared factors are.

In 1992 Bouchard published a further paper in which he estimated the heritabilities of the so-called "big five" personality traits.[9] These five were recognized in the 1980s by widespread consensus among personality theorists as the major dimensions of personality. Bouchard's estimates of their heritabilities are: extraversion (.49); neuroticism (.41); conscientiousness (.38); agreeableness (.35); and openness (.45). This analysis also showed that the shared or common family environment made no significant contribution to the variance in these personality traits.

WORK ATTITUDES

Bouchard and his group have also used their separated identical twins to investigate the heritability of various attitudes towards work. Twenty-three of their twins were given a questionnaire of work motivation.[10] The correlation between the twin pairs was .43 which, corrected for reliability, indicated a heritability of 68 percent. In a further study 34 of the separated twins were given a questionnaire of job satisfaction in which a correlation of .31 was obtained between the twin pairs, indicating a 31 percent heritability.[11] This correlation was not corrected for measurement unreliability.

Bouchard has never drawn any policy conclusions from his work on the heritability of intelligence and personality. He has never done any work on or offered any opinion on race differences, and he was a member of an American Psychological Association Task Force Report which met in 1995 and concluded that there is no direct evidence for a genetic basis for the racial differences in intelligence.[12] He has nevertheless been subjected to a certain amount of harassment from student activists at the University of Minnesota. His work has also been criticized in *Scientific American* by John Horgan,[13] who attempts to link Bouchard studies with the views of Hitler and the Nazis and argues that the work contains a hidden eugenic agenda, claims too far fetched to have any credibility.

Notes

1. Bouchard, T. J., Jr. & McGee, M. G. 1977. Sex differences in human spatial ability: Not an X-linked recessive gene effect. *Social Biology*. 24: 332-335.

2. Bouchard, T. J., Jr. 1996. IQ similarity in twins reared apart; findings and responses to critics. In R. Sternberg and C. Grigorenko (eds.) *Intelligence: Heredity and Environment*. New York: Cambridge Univ. Press.

3. Burt, C. L. 1966. The genetic determination of differences in intelligence. *British Journal of Psychology*. 57: 137-153.

4. Kamin, L. 1974. *The Science and Politics of IQ*. Potomac, MD.: Erlbaum.

5. Bouchard, T. J., Jr. & McGue, M. G. 1981. Familial studies of intelligence: a review. *Science*. 212: 1055-1059.

6. Bouchard, T. J., Jr. 1993. The genetic architecture of human intelligence. In P. A. Vernon (ed.) *Biological Approaches to the Study of Human Intelligence*. Norwood, NJ: Ablex.

7. Finkel, D., Pedersen, N. L., McGue, M. G. & McClearn, G. E. 1995. Heritability of cognitive abilities in adult twins: comparison of Minnesota and Swedish data. *Behavior Genetics*. 25: 421-431.

8. Tellegen, A. Lykken, D. T., Bouchard, T. J., Jr., Wilcox, K. J., Segal, N. L. & Rich, S. 1988. Personality similarity in twins reared apart and together. *Journal of Personality and Social Psychology*. 54: 1031-1039.

9. Bouchard, T. J., Jr. 1992. Genetic and environmental influences on adult personality: Evaluating the evidence. In Hettema, J. and Deary, I. (eds.) *Basic Issues in Personality: European and American Workshop on Biological and Social Approaches to Individuality*. Dordrecht: Kluwer.

10. Keller, L. M., Bouchard, T. J., Jr., Arvey, R. B., Segal, N. L. & Davis, R. V. 1992. Work values: Genetic and environmental influences. *Journal of Applied Psychology*. 77: 79-88.

11. Bouchard, T. J., Jr. 1989. Job satisfaction: Environmental and genetic components. *Journal of Applied Psychology.* 74: 187-192.

12. Neisser, U. 1996. Intelligence: knowns and unknowns. *American Psychologist.* 51: 77-101.

13. Horgan, J. 1993. Eugenics revisited. *Scientific American.* June: 92-100.

David T. Lykken, Ph.D.
University of Minnesota

Chapter 24

David T. Lykken

David T. Lykken (born 1928) has devoted most of his professional career to the study of the psychology of crime and psychopathic and sociopathic personalities, the underlying personality types responsible for crime. His research has included the neurophysiology of these psychopathological personality types, the genetic and environmental factors contributing to them, the psychology of the socialization of children, and the steps that should be taken by society to reduce the increases in crime and the underlying psychopathologies.

David T. Lykken was born in 1928 in Minnesota. His grandparents were Norwegian immigrants who had come to Minnesota as farmers in the late 19th century. Lykken's father was an engineer and inventor, and the holder of more than 50 patents. David Lykken was the youngest of five brothers, three of whom became engineers like their father, and Lykken himself entered the University of Minnesota in 1946 planning to major in chemical engineering. However, he took an introduction to psychology course and liked it, and this led him to take other courses in psychology. In his junior year, Lykken realized that he had already arrived at the cutting edge of this new science while, in chemistry, he had not progressed much beyond Lavoisier. Scientific psychology was still in its infancy, most of its seminal ideas yet to be formulated, and he believed that important discoveries were like sea shells on a virgin beach, just waiting to be picked up. He viewed psychology as crude, imprecise, largely atheoretical, a job for a rough carpenter rather than a cabinet maker, and he saw it as a good match for his talents and limitations. So he changed his major and then never looked back.

Lykken graduated B.A. in 1949 and stayed on at Minnesota for graduate work. He specialized in learning theory for his M.A. and then switched to clinical psychology. Although he had not yet written up his dissertation, those were simpler times, and Lykken received from the National Science Foundation a post-doctoral fellowship to spend 1954-1955 in London with Hans Eysenck. Returning to Minnesota, he was appointed assistant professor of psychiatry in 1957 and was promoted to tenure during 1959-1960 which he spent as a fellow at the Behavioral Sciences Center in Palo Alto. Lykken became a research professor of psychiatry in 1965, and then, in 1989, he moved to Minnesota's psychology department.

EARLY RESEARCH

Lykken's Ph.D research tested the theory that the source of primary (innate) psychopathy might be an innately low level of fearfulness and produced empirical results which supported this conclusion.[1] This result has been replicated and extended by others and is one of the most important discoveries ever made on the psychopathic personality. Lykken has synthesized and summarized work on the psychopathic personality in his 1995 book *The Antisocial Personalities*.[2] Lykken's early work was concerned with the nature and measurement of electrodermal reactions (skin sweating reactions), conditioning, and habituation in psychopaths and normals, and in the actuarial prediction of crime, personality measurement, psychological genetics, and gene-environment correlation, and he has devoted his career to the study of these phenomena, on which he has published more than 170 articles and book chapters.

POLYGRAPHIC INTERROGATION

One of Lykken's interests in his early career was in the lie detector. He constructed a short lie detector with the objective of attracting the interest of the prison inmates he was testing. In 1959 and 1960 he published two papers on what he called the Guilty Knowledge Test, a method of polygraphic interrogation that, unlike the lie test, has good scientific credentials.[3] These papers established a tradition for the *Journal of Applied Psychology* to publish research in forensic psychology. When the lie detection industry invaded the private sector in the 1970s and rapidly spread throughout American society, Lykken became the principal scientific critic of what he considered this misuse of applied psychophysiology. He published 12 journal articles on polygraphy, 6 chapters in edited volumes, and a book, *A Tremor in the Blood*,[4] still the only scientific monograph on this topic. He has been called upon to impeach lie detector evidence in state, federal, and military courts in more than 50 cases and to testify on the scientific status of the polygraph before a Canadian Royal Commission, a committee of the British Parliament, legislative committees of several states as well as three committees of the U.S. House and Senate. From 1970 until recently he estimates that he has spent 25 percent of his professional time in giving expert testimony relating to polygraphic interrogation.

GENETIC RESEARCH

Lykken was a student during the heyday of behaviorism and, like his mentors, developed an exaggerated notion of the ability of then current learning theories to account for the variation in human traits. Such Watsonian radical environmentalism was never the party line at Minnesota, however, and both Lykken and his teachers thought it perfectly acceptable to postulate a genetic temperamental factor in the psychopathic personality. Lykken started using twins as research subjects in 1969, and in 1979 Tom Bouchard asked him to collaborate on his extensive study of twins separated in infancy and reared apart. From this time onwards Lykken shifted his interests from psychophysiology to the mechanisms by which genetic differences become expressed in differences in complex psychological traits. With his colleagues Bouchard, M. McGue, and A. Tellegen, he has established the largest birth-record-based twin registry in the United States and, with W. Iacono, McGue, and Tellegen, he has begun a study of 11 and 17 year old twins and their families, 1,200 families in all, with plans to follow the 11 year olds through adolescence to age 20, and the 17 year olds to young adulthood.

EMERGENESIS

An important study of resting EEG (electroencephalographic) spectra in adult twins carried out by Lykken in collaboration with W. G. Iacono and A. Tellegen[5] showed that most monozygotic (MZ) twins' EEG spectra are as similar to those of their co-twins as they are to the first twins' spectra measured on a second occasion. The spectra of dizygotic (DZ) co-twins, however, are much less similar. This suggests that the genetic mechanisms involved are non-additive, i.e. do not involve the additive effect of a number of genes working through polygenic inheritance. Lykken later presented evidence of other traits on which MZ twins, including twins separated in infancy and reared apart, are very similar while DZ twins are little more alike than unrelated persons. This pattern of inheritance suggests that the genetic processes involved are not additive polygenes, because these would produce relatively high correlations of 0.5 between DZ twins, as well as between siblings. The correlations should be at least half as much as those obtained for MZ twins, but the actual correlations are very much lower than this.

To explain this phenomenon, Lykken proposed a genetic process the essence of which is that many psychological traits are determined by a particular pattern of genes rather than additively. The whole pattern has to be present for the trait to be expressed, and one missing gene will destroy the whole pattern. Thus one sibling could have the complete pattern, while the co-sibling might have all the required genes bar one. The sibling pairs would then be genetically similar, and psychologically different traits determined in this way would be strongly genetic in origin but would not tend to run in families. Emergenesis would explain the "black sheep" phenomenon, the sudden appearance of a psychopath in families that have been well socialized for generations. The black sheep happens to have a unique configuration of genes predisposing to psychopathic personality.[6] Further evidence for this theory including its application to the sudden appearance of exceptionally fast race horses and an illustration of how the occurrence of adaptive emergenic traits might be a plausible mechanism for rapid spurts of evolutionary change has been presented by Lykken and his colleagues, Bouchard, McGue, and Tellegen.[7]

Lykken believes that research in behavior genetics during the 1980s and 1990s has led to two possibly surprising conclusions. The first of these is that all psychological characteristics that have been measured have from 25 percent to 80 percent of their variation determined genetically. This is true for IQ and other aptitudes, traits of temperament and personality, interests, attitudes, and values, and even complex characteristics such as risk for divorce and the quality of happiness itself. [8]

The second and probably more surprising conclusion of genetic research is that being reared together in the same home by the same parents does not appear to have much or even any effect on making children more alike psychologically when they reach adulthood. This has been shown for unrelated adopted siblings, who have very little resemblance for personality traits, and it can be seen also in the fact that for many psychological traits DZ twins are less than half as similar as MZ twins, although shared home environment might be expected to make DZ similarity more than half of that found for MZs.

CRIME, PSYCHOPATHY, AND SOCIOPATHY

Lykken has been concerned about the rise of crime in the United States in the post World War II decades and has grappled with the problems of the causes of this increase and how society can deal with it. He believes that crimes are committed by psychopathic and sociopathic personalities, both of which are unsocialized to respect the law and the feelings of others. His distinction between psychopaths and sociopaths is that psychopaths have a strong genetic predisposition to fail to be socialized adequately and hence to become criminals; sociopaths are caused environmentally by incompetent parents who fail to socialize their children by training and example. Lykken likens the socializing of children to the training of dogs and draws on his personal experience of raising bull terriers, a breed initially selected for fighting ability and temperament.

Unusually for a psychologist, he believes that an occasional slap for misdemeanors is effective in the socialization of both dogs and children. He has presented an integrated account of his conclusions on these issues in his book *The Antisocial Personalities*.[9] Lykken notes that crime has increased about fivefold over the period 1960-1990, and considers that this must be largely due to environmental factors or, in his terminology, an increase in the prevalence of sociopathy rather than in psychopathy. He notes that about 70 percent of sociopaths, e.g., delinquents, school dropouts, and single teenage mothers, are brought up by single mothers, most of whom are living on welfare. He argues that an important constituent of child socialization is the presence in the family of a father who works and provides a role model for well socialized and law abiding behavior. Thus, he sees the rise in crime as closely associated with and partly caused by the breakdown of the nuclear two parent family.

Thus, Lykken regards parental socialization as an exception to the principle that the parent's style of child-rearing has little effect on the personality of the child. He believes that children reared together by immature, incompetent, or unsocialized parents tend to remain unsocialized themselves. For similar reasons we could suppose that children reared together by parents who are mute would tend to remain mute themselves; our human capacities for language and for socialization need to be elicited, shaped, and practiced with the help of speaking socialized adults. Children of mute parents residing in a community of speakers will learn language with the help of their peers and neighbors and, similarly, children of incompetent or unsocialized parents may acquire the rudiments of socialization through association with a socialized peer group. Unfortunately, many unsocialized parents reside in the subculture known as the underclass so that their children are doubly handicapped.

Lykken believes it is important to research more closely the socialization strategies of parents and their effectiveness in producing well socialized children. To do this he set up a "Parenting Project," which has been supported since 1993 by the Pioneer Fund and is a study of socialization in 300 pairs of 30 to 33 year old male twins from the Minnesota Twin Registry. Self-report measures of antisocial behavior, both in adolescence and later, obtained through questionnaires and interviews of the twins, their mothers and their spouses, plus questionnaire measures of religious commitment, empathy, altruism, and social responsibility have been obtained. Possible causal factors being investigated are the traits of aggressiveness, fearlessness, and stimulus-seeking, all known to be strongly genetic in origin and which are risk factors for criminality, and retrospective measures of the quality of parenting.

Although these twins have broadly middle class origins and few could be described as criminal, they show wide variation in the frequency and seriousness of the antisocial behaviors they acknowledge. Because the true underclass is not represented in this sample, however, it is anticipated that traits of innate temperament will account here for more of the variance in sociopathic and antisocial behavior than will differences in parental child rearing competence. That is, it is expected that one major finding of this study will be to confirm the hypothesis that genetic temperamental differences account for much of the variance in middle class criminality, and that the variance in the ability of middle class parents to socialize their children is too slight to explain much of the variance in the antisocial behaviors of those children during adolescence and young adulthood.

However, preliminary results, based on returns from 196 twins, indicate that the new measure of parental characteristics, as reported retrospectively by the young men, predicts antisocial behavior significantly better than a composite measure of crime relevant temperament (aggressiveness, stimulus seeking, impulsiveness, etc.) and better also than a quite detailed assessment of the degree of religious commitment, both when growing up and in adulthood. Other components of socialization in addition to avoidance of antisocial behavior — the traits of altruism, empathy, and responsibility — are strongly intercorrelated but only modestly related to poor socialization. It appears that one can be a frequent rule breaker and yet be altruistic and empathic and vice versa. The same parenting factor that best predicts antisocial behavior, however, is also the best predictor of this second facet of socialization.

Finally, it appears that parental strictness is correlated positively with the "good parent" factor in these preliminary results. Thus, the parents' child rearing skills may be more important in socializing middle class children than was initially supposed. Lykken believes that a similar study conducted on the true underclass of American society would show an even stronger effect of parental child rearing incompetence and a corresponding low heritability for poor socialization.

LICENSES FOR PARENTHOOD

Lykken's major solution to the growing crime rate is to introduce a system of licenses for parenthood. Because the psychopaths and sociopaths who commit crimes are largely the children of psychopaths and sociopaths, the objective should be to try to prevent children from being born into this situation. This would be done by requiring couples to obtain licenses for parenthood before they are permitted to have children. Lykken's idea is similar to that originally proposed by Sir Francis Galton in his eugenic utopia, but there is the difference that Galton was aiming to eliminate the birth of genetically at risk

babies, whereas Lykken is concerned to reduce the
birth of babies whose mothers are not capable of
rearing them competently.

The licenses for children proposed by Lykken
would only be granted to couples who are married,
financially independent, and have no serious
criminal convictions or psychiatric disorder. In
addition, applicants for a parental license would be
required to attend a course and pass an examination
on child rearing. He does not include mental
retardation as a reason for withholding a parental
license.

One of the problems in the parental licensing
proposal is to spell out what action would be taken
against women who committed the new crime of
producing unlicensed babies, and against the fathers
of these babies, if these could be identified. Since
about a third of the babies born in the United States
in the last decade of the 20th century were
unplanned, the new crime would no doubt be
committed on an extensive scale. To deal with this
issue, Lykken proposes that unlicensed babies
should be taken away from the mother and reared
by foster parents or in boarding schools staffed by
no-nonsense noncommissioned army officers and
similar well-socialized persons. Any woman
producing a second unlicensed baby would be
required to receive a long-lasting contraceptive.

Lykken does not advocate parental licensing
on genetic or eugenic grounds. His concern is rather
to prevent motherhood by those lacking the skills
necessary to socialize their children properly.

NOTES

1. Lykken, D. T. 1957. A study of anxiety in the sociopathic personality. *Journal of Abnormal and Social Psychology*. 55: 6-10.

2. Lykken, D. T. 1995. *The Antisocial Personalities*. Hillsdale, NJ: Erlbaum.

3. Lykken, D. T. 1959. The GSR in the detection of guilt. *Journal of Applied Psychology*. 43: 385-388; Lykken, D. T. 1960. The validity of the guilty knowledge technique: the effects of faking. *Journal of Applied Psychology*. 44: 258-262

4. Lykken, D. T. 1980. *A Tremor in the Blood: Uses and Abuses of the Lie Detector*. New York: McGraw-Hill.

5. Lykken, D. T., Tellegen, A. & Iacono, W. G. 1982. EEG spectra in twins: evidence for a neglected mechanism of genetic determination. *Physiological Psychology*. 10: 60-65.

6. Lykken, D. T. 1982. Research with twins: the concept of emergenesis. *Psychophysiology*. 19: 361-373.

7. Lykken, D. T., Bouchard, T. J., Jr., McGue, M. & Tellegen, A. 1992. Emergensis: genetic traits that do not run in families. *American Psychologist*. 47: 1565-1577.

8. Lykken, D. T., Bouchard, T. J., Jr., McGue, M. & Tellegen, A. 1993. Heritability of interests: a twin study. *Journal of Applied Psychology*. 78: 649-661.

9. Lykken, D. T. 1995. *Op. cit.*

Joseph M. Horn, Ph.D.
University of Texas at Austin

Chapter 25

Joseph M. Horn

Joseph M. Horn (born 1940) is a psychologist whose principal interests lie in intelligence and personality, and their genetic and environmental determinants. His work on the Texas Adoption Project has provided some of the most persuasive evidence that genetic factors exert strong effects on both intelligence and personality.

Joseph M. Horn was born in 1940 in Stillwater, Oklahoma. In 1963 he graduated from Oklahoma State University with a degree in chemical engineering and worked for Du Pont for a year before going into the U.S. Army for two years, some of which time he spent in Vietnam. His army responsibilities allowed enough time to take psychology courses through American University, and he became so interested in psychology that after leaving the army he entered the graduate school in psychology at the University of Minnesota, from which he obtained an M.A. in 1967 and a Ph.D. in

1969. Horn's research for his doctorate was in the field of behavior genetics and was concerned with the heritability of aggression in mice.

In 1969 Horn was appointed assistant professor of psychology at the University of Texas. He has remained at Texas for the rest of his career and was appointed full professor in 1984. On taking up his position in Texas, he shifted his interest from behavior genetics in mice to that in humans and decided to tackle this by a long term study of adopted children and the degree to which they resemble their biological and adoptive parents. The logic of this research design is that if adopted children resemble their biological parents for a trait, there must be some element of genetic transmission. This idea gave birth to the Texas Adoption Project. The Project began in 1971 with the discovery of an adoption agency in Texas which had administered a number of psychological tests to unmarried mothers-to-be and was willing to allow Horn access to these records and to the identities of their babies and their adoptive parents. The first phase of this research program was funded by the National Institute of Mental Health. The Project has received support from the Pioneer Fund from 1975 onwards.

The project has involved 300 Texas families that had adopted one or more children from a home for unwed mothers. The total number of individuals tested in the project was approximately 1,230, who included the adopted children, the adoptive mothers and fathers, the biological mothers, and other biological and adopted children in the adoptive families. In the first phase of the study, the adopted children were between 3 and 14 years old; 10 years later approximately 258 adopted children in 181 families were located and tested again, together with 93 biological children of the adoptive parents. These data have been analyzed in various ways to provide estimates of the contribution of genetic and environmental factors to intelligence and personality. The project has been run by Horn in collaboration with his University of Texas colleagues John Loehlin and the late Lee Willerman.

PSYCHOPATHIC PERSONALITY IN UNWED MOTHERS

One of the first publications arising from the project concerned the high incidence of psychopathic tendencies in the unwed mothers.[1] These mothers were tested with the Minnesota Multiphasic Personality Inventory, which measures 12 psychopathological personality traits. The study compared the scores of the 363 unmarried mothers with those of a large sample of 18 year old young women and a smaller sample of young married women. The most striking result of the study was the high level of psychopathic personality disorder among the unwed pregnant women. Their average psychopathic deviate scores were approximately one standard deviation higher than that of the control groups of young women. They also scored higher on the schizophrenia, paranoia, psychasthenia, and hypomania scales.

Horn and his colleagues argued that this showed that adopted children were likely to inherit psychopathological tendencies from their mothers. They noted that the incidence of psychopathological conditions was approximately four times greater among adopted children than among children reared by their biological parents. Some psychologists have attributed this greater incidence to the psychologically disturbing effects of the experience of being adopted, but Horn and his colleagues argued that it is more probably due to genetic transmission.

Horn's team also argued that the high incidence of psychopathic and other psychopathological tendencies in unwed mothers leads to an underestimation of the extent of the heritability of psychopathological disorders from adoption studies. The methodology employed in these studies has been to compare the incidence of psychopathological disorders among the adopted children of affected biological parents as compared with those of control (unaffected) biological parents. If the incidence is greater among the adopted children of the affected parents, some genetic determination of the disorder is indicated. They argued that significant numbers of the control parents may also have undiagnosed psychopathological tendencies. This would reduce the difference between the two groups and would bias studies of this kind towards an underestimation of genetic effects.

HERITABILITY OF INTELLIGENCE

As the Texas Adoption Project has progressed, the adopted children have been tested for both intelligence and personality during childhood and adolescence and in their early adult years. This has made it possible to calculate heritabilities at different ages, and to elucidate the nature of the environmental effects on intelligence.

So far as intelligence in concerned, an analysis of the data published in 1983 showed that the adopted children resembled their biological mothers more closely than their adoptive mothers, the two correlations being 0.28 and 0.15, respectively.[2] This result indicates that both genes and the environment affect the development of children's intelligence, and that genetic effects are about twice as strong as environmental effects. Horn and his colleagues made a number of calculations of the heritability of intelligence by breaking the sample down into males and females and different age and socioeconomic status categories. They estimated narrow heritabilities (those attributable only to additive genes) for their various subsamples as ranging between 0.22 and 0.62. They estimated the influence of common family environment at only 0.18.[3] This means that a significant impact on intelligence must be made by environmental factors that affect one sibling but not the other, i.e., unique environmental effects, consisting of such things as a debilitating illness contracted by only one of the siblings.

The adopted children were retested in adolescence and early adulthood when they were aged between 13 and 24 years. At this stage the correlation between the intelligence of the adopted children and those of their biological mothers had increased slightly to 0.26, while the correlation between the intelligence of the adopted children and their adoptive parents had declined to 0.05.[4] This result indicates that among adolescents and young adults genetic influences on intelligence are about five times as strong as family environmental influences. It shows also that genetic influences are more powerful on the intelligence of adolescents and young adults than they are on young children. To put this another way, family influences on intelligence operate during early childhood, but fade to insignificance by adolescence and early adulthood.

This inference is supported by the pattern of the correlation between pairs of unrelated children adopted by the same parents. Among young children these unrelated pairs showed a correlation of 0.11, but in adolescence and early adulthood this correlation had dropped to -0.09.[5] This suggests that there are some short term common family effects on the children's intelligence which bring the unrelated young siblings' IQ into positive correlation. These effects are probably things like the richness of the vocabulary used by the parents, the provision of cognitively-stimulating games and so forth. However, these influences evidently have no long term effect, since the correlation has disappeared when the siblings are adults. This result is contrary to the theories often advanced about the beneficial effects of cognitive stimulation in the development of children's intelligence and suggests that cognitive stimulation has no long term value.

GENETICS OF SPECIFIC COGNITIVE ABILITIES

By around 1990 so much evidence had accumulated to show that general intelligence has a high heritability that the issue could be regarded as settled. However, there was still a question about whether the specific cognitive abilities, such as spatial, verbal, reasoning, and so forth, are under some genetic control over and above the genetic contribution to general intelligence. In 1994 Horn analyzed the data from the Texas Adoption Study to elucidate this problem. The results showed that the spatial and perceptual speed abilities do have a genetic component independent of general intelligence, although this was not the case for verbal ability.[6]

HERITABILITY OF PERSONALITY

Horn and his colleagues have collected data on personality in the Texas Adoption Project and used it to analyze heritabilities and environmental effects. Psychopathic tendencies were measured using the psychopathic scale of the Minnesota Multiphasic Personality Inventory in 138 adopted children and their biological and adoptive mothers.[7] The results showed that the psychopathic deviate scores of the biological mothers were correlated with the scores of the children approximately three times as strongly as the scores of the adoptive mothers. This indicates that the genetic transmission of psychopathic personality from biological mothers to their children is about three times as powerful as the environmental transmission from adoptive mothers to their adoptive children. Horn has also examined the strength of the association between the psychopathic deviate scores of pairs of unrelated adopted siblings. Among adolescents and young adults the correlation was -.09 or effectively zero. This indicates that common family rearing practices, discipline, example, and experiences have no long term impact on the extent to which adolescents and young adults develop psychopathic tendencies.[8]

Horn's team has also examined the question of whether siblings have an effect on each other's personalities. They have shown that unrelated individuals reared together do not in general become more similar over time.[9] For some traits they even become more dissimilar. For instance, as one sibling becomes more introverted, the other becomes more extroverted. This reinforces the conclusion that family influences on the development of personality are weak. The major influence on personality appears to be unshared environmental effects. Furthermore, an analysis of the changes in personality over ten years of childhood and adolescence indicated a shift towards poorer socialization among the adopted children, possibly due to increasingly strong genetic effects transmitted from their somewhat psychopathic biological mothers.

THE INTEGRITY OF THE MEDIA

All serious scholars who work on the genetics of intelligence have concluded that heredity makes an important contribution. Horn himself, taking into account the evidence from twin studies in addition to his own data on adopted children, estimates the heritability at between 60 and 80 percent.[10] Nevertheless, in the last quarter of the 20th century many journalists have disputed the evidence for the heritability of intelligence and the validity of the concept. A typical instance of the widespread journalistic misrepresentation on this issue occurred in 1975 when Dan Rather of CBS TV news presented a program called "The IQ Myth" in which he asserted that there is no evidence that individual differences in IQs are affected by genetic factors and that intelligence tests are not valid measures of cognitive ability. Horn wrote an article refuting these assertions.[11] He described the program as "an intellectual disgrace" and suggested that the researchers assembling the material for the program had received "some bad advice from a small number of politically motivated psychologists". [12]

Horn was correct in stating that there is a small number of politically motivated academics who consistently mislead technically unsophisticated journalists on these issues and that there is a significant number of journalists who present these minority views as if they represent the consensus position.

Notes

1. Horn, J. M., Green, M., Carney, R. & Erickson, M. T. 1975. Bias against genetic hypotheses in adoption studies. *Archives of General Psychiatry.* 32: 1365-1367.

2. Horn, J. M. 1983. The Texas Adoption Project: adopted children and their intellectual resemblance to biological and adoptive parents. *Child Development.* 54: 268-275.

3. Horn, J. M., Loehlin, J. C. & Willerman, L. 1982. Aspects of the inheritance of intellectual abilities. *Behavior Genetics.* 12: 479-516.

4. Loehlin, J. L., Willerman, L. & Horn, J. M. 1988. Human behavior genetics. *Annual Review of Psychology.* 29: 101-133.

5. Loehlin, J. L., Horn, J. M. & Willerman, L. 1989. Modeling IQ change: evidence from the Texas Adoption Project. *Child Development.* 60: 993-1004.

6. Loehlin, J. L., Horn, J. M. & Willerman, L. 1994. Differential inheritance of mental abilities in the Texas Adoption Project. *Intelligence.* 19: 325-336.

7. Willerman, L., Loehlin, J. C. & Horn, J. M. 1992. An adoption and a cross fostering study of the Minnesota Multiphasic Personality Psychopathic Deviate Scale. *Behavior Genetics.* 25: 515-529.

8. Loehlin, J. L., Willerman, L. & Horn, J. M. 1987. Personality resemblance in adoptive families: a 10 year follow up. *Journal of Personality and Social Psychology.* 53: 961-969.

9. Loehlin, J. L., Horn, J. M. & Willerman, L. 1990. Heredity, environment and personality change: evidence from the Texas Adoption Study. *Journal of Personality.* 58: 221-243.

10. Loehlin, J. L., Willerman, L. & Horn, J. M. 1988. *Op. cit.*

11. Horn, J. M. 1975. The IQ myth revisited. *Texas Psychologist.* 27: 19-21.

12. *Ibid.* 19.

Chapter 26

Hoben Thomas

Hoben Thomas (born 1936) has devoted his career to the study of child and developmental psychology, individual difference personnel selection and sex differences in intelligence. One of his principal interests is the construction of a mathematical model to explain the superior average performance of males on spatial and math abilities in terms of X-linked recessive genes.

Hoben Thomas was born in 1936 and took his first degree in psychology at the University of California at Santa Barbara in 1958. He obtained an M.A. at Pepperdine College in 1959 and a Ph.D. at Claremont University in 1963. His first appointment was at Fresno State College, where he worked from 1964-66. In 1966 he joined the psychology department at Pennsylvania State University, where he has spent the remainder of his career.

SEX DIFFERENCES IN SPATIAL ABILITY

One of Thomas's major interests has been in the problem of sex differences in spatial ability. It has long been known that males tend to perform better than females on tests of spatial ability, particularly in the ability to visualize how an object would appear if it is moved or rotated. In 1961 R. E. Stafford proposed that this kind of rotational spatial ability might be determined by an X-linked recessive gene and that this would account for the male superiority on these tasks.[1]

Characteristics determined by X-linked recessive genes appear in males, but not in females. Color blindness and hemophilia are two well known conditions that are determined by X-linked recessive genes and are more common in males than in females. Stafford's hypothesis was that spatial rotational visualization is inherited in the same way. The reason that a recessive linked characteristic is more common in males than in females is that males have one X chromosome and one Y chromosome, whereas females have two X chromosomes. Males with one recessive allele in their X chromosome express the condition, whereas in females the condition is more typically suppressed by the presence of a normal dominant allele on their other X-chromosome.

A number of studies have been carried out designed to test the X-linked recessive gene theory of spatial rotational ability, and in general the results have been considered to be negative.[2] However, Thomas has constructed a mathematical model consistent with the theory. The X-linked recessive gene theory implies that the proportion of males with the ability should be the square of the proportion of females, and Thomas has shown that in several data sets this is approximately the case.[3]

In a further paper Thomas argued that the failure of the ordering of familial test score correlations (mother-daughter, mother-son, father-daughter, father-son) for spatial abilities to support the X-linked recessive gene hypothesis does not invalidate the hypothesis. He argued that these correlations are determined partly by environmental effects which could distort the magnitudes of the correlations predicted from the X-linked recessive gene theory, and that the rejection of the hypothesis by a number of behavior geneticists is premature.[4]

In a subsequent paper Thomas has adduced further arguments in favor of the X-linked recessive gene hypothesis, such as that females should show greater variability in the ability and that spatial rotational ability should be distributed as two normal distributions rather than one.[5] The conclusion of Thomas's work is that the X-linked recessive gene theory is still a possible hypothesis for the male advantage on spatial rotational ability.

SEX DIFFERENCES IN MATHEMATICAL ABILITY

Hoben Thomas has also worked on the problem of sex differences in mathematical ability. It has been found in a number of studies that males tend to obtain higher average scores on mathematical tests than females. This difference between the sexes is present among normal representative samples and also among samples of the mathematically gifted.

The largest study of the mathematically gifted is that of Camilla Benbow and Julian Stanley who have tested approximately 65,000 American adolescents for high mathematical aptitude.[6] They selected the top scoring 3 or so percent and gave them the math section of the Scholastic Aptitude Test (SAT-m). They found five significant facts. First, the SAT-m mean for boys was higher than that for girls; second, there were more mathematically precocious boys than girls; third, at a relatively modest level of mathematical ability, represented by a SAT-m score above 420, the ratio of boys to girls was about 1.5:1, but at a high level of ability, represented by a SAT-m score of 700, the ratio of boys to girls was much larger at approximately 13:1; fourth, the male variance is greater than the female; and fifth, there was no difference in the mean scores on the verbal SAT among the sample of mathematically precocious adolescents. Neither Stanley nor Benbow, nor anyone else, has been able to provide a satisfactory explanation for these findings.

Many psychologists assert that the sex differences in mathematical ability is brought about by environmental conditioning, such as the social expectation that girls are not as good as boys at math. Hoben rejects explanations of this kind and proposes that the data can be explained by positing an X-linked recessive gene facilitating mathematical ability. The gene would operate in the same way as the postulated X-linked recessive gene for spatial ability, producing many more males with the ability than females.[7]

In 1995 the Pioneer Fund made a grant to Hoben Thomas to carry out research on women with high mathematical ability using a sample of female math professors, and the possible genetic transmission of their math abilities to their children. At the time of writing the research has not been completed.

NOTES

1. Stafford, R. E. 1961. Sex differences in spatial visualization as evidence of sex-linked inheritance. *Perceptual and Motor Skills*. 13: 428.

2. Plomin, R. DeFries, J. C. & McClearn, G. E. 1989. *Behavioral Genetics*. New York, NY: Freeman.

3. Thomas, H. & Jamison, W. 1981. A test of the X-linked genetic hypothesis for sex differences on Piaget's water level task. *Development Review*. 1: 274-283.

4. Thomas, H. 1983. Familial correlational analyses, sex differences and the X-linked gene hypothesis. *Psychological Bulletin*. 93: 427-440.

5. Thomas, H. & Krail, H. 1991. Sex differences in speed of mental rotation and the X-linked hypothesis. *Intelligence*. 15: 1732.

6. Benbow, C. & Stanley, J. 1980. Sex differences in mathematical ability: fact or artifact? *Science*. 210: 1262-1264.

7. Thomas, H. 1985. A theory of high mathematical aptitude. *Journal of Mathematical Psychology*. 29: 231-242.

Chapter 27

Brunetto Chiarelli

Brunetto Chiarelli (born 1934) is one of the foremost Italian anthropologists. He has written extensively on the evolution of primates and he specializes in genetic and chromosomal variations in monkeys, apes, and humans.

Brunetto Chiarelli was born in Florence in 1934. He studied anthropology, biology and genetics at the University of Florence for his first degree. He obtained a Ph.D. in anthropology from the University in 1957 and in genetics in 1960. From 1958-1959 Chiarelli worked as assistant professor at the Institute of Anatomy in Florence. In 1960 he moved to the University of Pavia to an assistant professorship at the Institute of Genetics. From 1962-1979 he worked initially as associate professor and later as full professor at the University of Turin. In 1979 he returned to Florence as professor of Anthropology and director of the Institute of Anthropology at the University of Florence.

Chiarelli has published more than 200 papers, principally in the fields of comparative genetics, cytogenetics, the taxonomy of primates, the biology of human populations, and bioethics. He has also written eight books, including *The Origin of Man; The Evolution of the Primates*; and *Global Bioethics*.[1] He has recently written an account in Italian of the evolution of intelligence from reptiles through mammals to humans, and of the evolution of human racial differences in intelligence, in which he has presented the theories of Jensen and Lynn to Italian scholars.[2] Chiarelli has received support for research from the Pioneer Fund from 1990 onwards.

GENETICS OF ISOLATED POPULATIONS IN ITALY

The Pioneer Fund's first grant to Chiarelli was made for a study of the genetic characteristics of two isolated populations in the Garfagnand Valley in Tuscany. This research studied variability of DA/DAPI and C chromosomal heteromorphic sites in a sample of 136 unrelated individuals, comparing two subsamples from the middle and upper valley respectively. The variations of DA/DAPI fluorescence at one specific site demonstrated an excess of homomorphic individuals in the upper valley, which could be related to the mating structure of the population of that area. These markers are a powerful tool in clinical diagnosis of the genetic relationship of populations, since they are Mendelian in their pattern of inheritance and little subject to mutation. As a result they have been used for paternity exclusion purposes, and this study further explored their utility as a tool for determining the degree of biological relationship between different populations. The study was conducted by Sergio Tofanelli and Mara Agostini of the University of Florence, Roscoe Stanyon of the University of Genoa, and Marcello Giovanni Franceschi and Giorgio Paoli of the University of Pisa under Chiarelli's overall direction.[3]

BIODEMOGRAPHY AND HUMAN EVOLUTION

Chiarelli also received support from the Pioneer Fund to hold a conference on Biodemography and Human Evolution in Florence in April, 1995. The objective of the conference was to bring together a number of international scholars to exchange ideas on the impact of humans on the environment and the effects of the pressure exerted by world population growth in the 20th century on the global ecology. The conference covered five themes. These were the impact of early man on the natural environment; paleodemography; the biodemography of historical populations; the present state of world population; and prospects for the future. The international authorities at the conference included Robert Foley of the University of Cambridge on the impact of the early dispersion of man on the global environment; F. Coppa of La Sapienza University, Rome, on paleodata of different geographical areas; E. Lucchetti of the University of Parma on the biodemographic structure of populations of Western Europe; E. Rabino Massa of the University of Turin on the biodemographic behavior of mountain populations; J. Hutchinson of the University of Texas on gender as a starting point for the construction of social subjects; P. J. Nas of the University of Leiden on urban anthropology; B. Barich of La Sapienza University, Rome, on the dynamics of population movements in response to climatic changes in Africa; C. Blanc Stanton of Columbia University on the global problem of children in distress; K. Omoto of Japan on the survival of the Ainu on the northern Japanese island of Hokkaido; G. Vona of the University of Cagliari on the peopling of Sardinia; P. Laureano of UNESCO on the proper use of natural resources; M. Milani Comparetti on the ethical issues involved in the use of birth control to limit population growth; and A. Camperio Ciani of the University of Padua on deforestation.

One of the most important papers presented at the conference was Roger Pearson's "Biological diversity and ethnic identity." Pearson presented a theory first advanced by George Gaylord Simpson in 1949. It proposed that humans have developed an instinctive sense of their own ethnic identity and prejudice against other ethnic and racial groups. The reason this has evolved is that it restrains cross breeding. It has also led to the evolution of altruism, the willingness of the individual to help other members of his own group and if necessary to sacrifice his own life for its benefit.

Pearson argued that in the modern world the greater communications and interaction through travel between different races and ethnic groups has led to interbreeding and a breakdown of human instinctive in-group altruism. Many people have replaced these with feelings of altruism towards the world population. Indeed, it is not uncommon for some people today to feel greater sympathy for the sufferings of distant groups than for those who live next door. Pearson doubts whether this global altruism will persist and spread indefinitely. Rather, he concludes that it is more likely that some groups will continue to preserve their sense of ethnic identity and restrict their altruism to their own group.

The conference closed with a round table discussion in the main University Hall of Florence. W. Hern (University of Colorado) explained his theory on the relationship between the impact of human culture on the environment and the impact of carcinoma on humans. E. Chiavacci (University of Florence) spoke on "[t]he point of view of the Catholic Church on population increase and control," outlined the position of the Church on the birth control issue and explained that the Church refuses every means of fertility control based on constraint, but realizes that nowadays couples have to be sexually responsible, in harmony with changes occurring in society. C. Marchetti (IASA, Vienna) expressed his faith in sustainable growth of human populations, due to the increased power of technology. Livi Bacci (University of Florence) spoke about the demographic trends of the next 50 years and argued that human population growth is now becoming a danger to the welfare of human beings, requiring a policy of fertility control not based on constraint but on values.

OLFACTORY VARIABILITY IN HUMAN POPULATIONS

In 1996 the Pioneer Fund made a further grant to Chiarelli to support a research project to be carried out by Francesco Scalfari on olfactory variability in human populations. The general approach of the project is that anthropologists and sociologists have given little attention to the human sense of smell, regarding it as of little importance for humans as compared with vision and hearing. Scalfari believes that the devaluation of the sense of smell has become particularly pronounced in economically developed western cultures. Smell is more important among non-western peoples in Africa, Oceania, and Latin America, where animals, plants, and people are identified by their smell to a much greater extent. The object of the research is to identify the biochemical bases of the sense of smell. These are believed to a be a certain class of proteins, some of which are known as Odorant Binding Proteins (OBPs). Two classes of these OBPs have been found in porcupines, deer, and rabbits. The research project aims to discover more about these proteins in humans and their variability in different human populations. The project promises to make an important contribution to existing knowledge of human genetic diversity.

In addition to his research and writing, Chiarelli acted as editor of the *Journal of Human*

Evolution from 1972 to 1985; of the *Antropologia Contemporanea* from 1978 to 1996; of *Human Evolution* from 1986 to 1996; of the *International Journal of Anthropology* from 1986 to 1996; and of *Global Bioethics* from 1990-1996. He has been a member of the advisory editorial board of *Mankind Quarterly* since 1988. He has also edited a number of books on anthropology and primate and human evolution.[4] In 1995 he served as Italian representative for the UNESCO Declaration against Racism, Violence and Discrimination.

The Science of Human Diversity
A History of the Pioneer Fund

NOTES

1. Chiarelli, B. 1973. *The Evolution of the Primates: An Introduction to the Biology of Man.* London: Academic Press; Chiarelli, B. 1978. *L'Origine dell'Uomo.* Bari: Laterza; Chiarelli, B. 1993. *Bioetica Globale.* Florence: Angelo Pontecorboli.

2. Chiarelli, B. 1995. Le basi evolutive dell' intelligenza. *Anthropologia Contemporanea.* 18: 97-117.

3. Tofanelli, S., Stanyon, R., Agostini, M., Francheschi, M. G. & Paoli, G. 1993. Variability of DA/DAPI and C heterochromatin regions: a population study. *Human Biology.* 65: 635-646.

4. Chiarelli, B. (ed.) 1991. *Language Origin: A Multidisciplinary Approach.* Dordrecht: Kluwer; Chiarelli, B. (ed.) 1992. *Proceedings of the VIIth Conference of the International Primatological Society.* Berlin: Springer-Verlag; Chiarelli, B. 1993. Chiarelli, B. (ed.) 1994. *The Ethological Roots of Culture.* Dordrecht: Kluwer.

Chapter 28

Medical Genetics

The Pioneer Fund has supported a number of studies of medical genetics and genetically based disorders, including sickle cell anemia, cancer, Tay-Sachs disease, hemophilia, schizophrenia, and inbreeding depression, as well as discussions of genetics, eugenics, and public policy. The Pioneer Fund joined the Hoover Institution in supporting the volume, *Evolution and Human Values*[1] edited by Robert Wesson and Patricia A. Williams, which examines the relationship between evolutionary theory, genetics, eugenics, and ethics. It contains a chapter, "Intelligence and Social Policy," by the late Richard Herrnstein, co-author of the best-selling *The Bell Curve*.[2] These projects are described in this chapter.

In 1978 and 1979 the Pioneer Fund made grants to the Sickle Cell Disease Foundation of Greater New York to enable it to hold courses on the disease for physicians, nurses, health educators, and social medical workers. Sickle cell anemia is a genetic disease present in a significant number of blacks in Africa, the United States, and elsewhere in the world. The disease is caused by a single recessive gene which is carried by approximately 18 percent of blacks in Central Africa and about 8 or 9 percent of blacks in the United States. A single recessive causes mild anemia but it is not strongly debilitating. A double recessive is present in approximately 1 percent of blacks in Africa and .04 percent of blacks in the United States and causes severe anemia.

The inheritance of the double recessive for the disease can be diagnosed in the fetus by the technique of amniocentesis and an affected fetus can be aborted. Attempts have been made to educate the black American population to the possibility that they may be carriers of the recessive and, if they mate with another carrier, at risk of having a child with the double recessive. If these educational efforts were successful an advance would be made toward the elimination of the disease. So far, relatively little has been done in this regard.[3]

THE GENETICS OF EYE CANCERS

In 1984 the Pioneer Fund made a grant to the New York Hospital-Cornell Medical Center in support of its research on inherited forms of eye cancer. The most common of these is retinoblastoma, a cancerous tumor of the retina developing in young children and invariably fatal if untreated. From the middle years of the century, the disease has been treated by surgical removal of the eyes. Unhappily, affected individuals are at greater than normal risk of developing other cancers later in life. Retinoblastoma is caused by a dominant gene, the effect of which is that if an affected individual mates with someone free of the disease, half the children inherit the gene and the disease. The development of the surgical treatment for the disease has had the effect that affected individuals who would previously have died in childhood are now surviving to adulthood and, in some cases, having children and perpetuating the presence of the gene in the population. Thus, it is an example of the dysgenic effects of medical advances in the treatment of individuals with genetic diseases who are enabled to survive to adulthood, have children, and transmit the adverse gene to future generations. Further research may in time make it possible to counteract this dysgenic effect, perhaps by removing the gene for the disease and inserting a healthy gene. The Pioneer Fund's grant was intended to promote progress on this front.

TAY-SACHS DISEASE

In 1984 the Pioneer Fund made a grant to the Eunice Kennedy Shriver Center for Mental Retardation in Waltham, Massachusetts, to support the research program on Tay-Sachs disease. This disease is common in Ashkenazi Jews, among whom about 1 in 30 is a carrier of the recessive gene for the disease, compared with about 1 in 300 among non-Jewish Caucasians. Because the gene for the disease is recessive it does no harm in carriers who have one copy of the gene, but if two carriers have children, an average of one in four of them inherits two copies of the gene and has the disease.

Tay-Sachs is an incurable degenerative disease and affected babies invariably die in early childhood. Since around 1970 mass screening programs in the American Jewish population have identified carriers of the gene. The result of this has been that if two identified carriers marry and the wife becomes pregnant, they have in the great majority of cases had prenatal tests carried out on the fetus to ascertain whether it has the lethal double recessive and, where a positive diagnosis has been made, had the fetus aborted. This program has virtually eliminated the birth incidence of the disease from the American Jewish population and has been one of the great eugenic successes of 20th century medicine, although it is not presented as such in the Shriver Center's promotional literature.

GENETICS OF SCHIZOPHRENIA

In 1989 the Pioneer Fund made a grant to Nechama S. Kosower, head of the genetics department at Tel Aviv University, for research on the genetics of schizophrenia. It has been known since the 1940s that identical twins are more similar (concordant) for schizophrenia than non-identicals, and therefore that genetic factors are involved. The objective of Kosower's research was to attempt to discover the location of the gene contributing to the disease.

The results of the work were published in 1995 and consisted of the discovery that schizophrenics differ from normal individuals in the size of a region on the first chromosome.[4] Technically, this is the heterochromatic region containing 1qH c-brand variants. The research found that the 1qH variants were smaller in schizophrenics than in normals. It was also found that schizophrenics have a greater frequency of the Duffy blood group. It was noted that this region contains the locus of the D5 dopamine receptor pseudo-gene, consistent with the theory that anomalies in dopamine production may be involved in schizophrenia.

HEMOPHILIA RESEARCH

In 1989 the Pioneer Fund awarded a grant to the National Hemophilia Foundation in New York City. Hemophilia is a genetic disorder of a failure of the blood clotting mechanism. About two thirds of cases are caused by an X-linked recessive gene which is transmitted by female carriers to their sons. The remaining one third or so of cases arise through spontaneous mutation. Bleeding can be stopped by injection of the protein factor in which hemophiliacs are deficient.

GENETIC RESEARCH IN SAO PAULO, BRAZIL

In 1992 the Pioneer Fund made a grant to Mayana Zatz, associate professor of genetics at the Center of Research on Human Variability and

Genetic Diseases at the University of Sao Paulo in Brazil to support a project to establish a DNA repository of permanent cell lines from patients affected by hereditary diseases and their families. This is intended to facilitate research into the heritability of various diseases and the identification of carriers or likely carriers of genes for the diseases. Once family pedigrees of genetic disorders have been mapped, it becomes possible to work out the mode of inheritance of the disorder and to assess the risks to family members of having an affected child. This enables individuals at risk to have a fetus tested for the presence of the disorder and aborted in cases where a positive diagnosis is made. A program of this kind can secure a considerable eugenic advance, as in the success of the virtual elimination of Tay-Sachs disease among the American Jewish population.

INBREEDING IN DAGHESTAN

In 1996 the Pioneer Fund made a grant to Helmut Nyborg of the University of Aarhus in Denmark to study the effects of inbreeding in Daghestan in the Northern Caucasus of the former Soviet Union. This remote mountainous region has been peopled by isolated populations for many generations and the populations have consequently become inbred.

A long term project studying the possible adverse effects of inbreeding among these populations has been made by Zazima Lulayeva of the Russian Academy of Science in Moscow. She has taken blood samples from the populations and collected a variety of psychological and neurophysiological data. The economic problems in Russia in 1995-1996 put the research program in jeopardy. The grant from the Pioneer Fund has made it possible to transfer the frozen blood samples and other data to Nyborg's laboratory in Aarhus where the DNA extraction, blood assays, and analysis of the psychological and neurophysiological data are being carried out.

NOTES

1. Wesson, R. and Williams, P. A. (eds.) 1988. *Evolution and Human Values*. Stanford, CA: Hoover.

2. Herrnstein, R. J. and Murray, C. 1994. *The Bell Curve: Intelligence and Class Structure in American Life*. New York, NY: Free Press.

3. Duster, T. 1990. *Backdoor to Eugenics*. New York, NY: Routledge.

4. Kosower, N. S., et al. 1995. Constitutive heterochromatin of chromosome 1 and Duffy blood group alleles in schizophrenia. *American Journal of Medical Genetics*. 60: 133-138.

PART V:
Race Differences

J. Philippe Rushton, Ph.D., D.Sc.
University of Western Ontario

Chapter 29

J. Philippe Rushton

J. Philippe Rushton (born 1943) began his professional career by working on altruism from an environmentalist perspective, but after some years shifted his viewpoint towards genetic and sociobiological explanations of this and other forms of human behavior. His most important work has been his Genetic Similarity Theory (an updated version of the theory of ethnocentrism), and his r-K theory of race differences (which proposes that a large number of Mongoloid-Caucasoid-Negroid behavioral differences can be understood in terms of different biological life history strategies for reproduction and survival).

John Philippe (Phil) Rushton was born in 1943 in Bournemouth, England, where his father ran a business as a building contractor. His mother was French and responsible for bestowing on him his middle name. Rushton was educated at Birkbeck College of the University of London, where he obtained his B.Sc. in psychology in 1970. He moved to the London School of Economics, from which he obtained his Ph.D. in 1973 for work on the development of altruism in children. He spent 1973-1974 continuing this work at the University of Oxford. In 1974 he emigrated to Canada and for the next several years continued his research on the development of altruism and other forms of moral behavior. He taught at York University (1974-1976) and at the University of Toronto (1976-1977). In 1977 he moved to the University of Western Ontario where he was made full professor in 1985.

STUDIES OF ALTRUISM

Rushton spent the first 10 or so years of his professional life working on the development of the human propensity to help others, which is known as pro-social behavior or altruism. The widespread presence of altruism raises the question of why people so frequently assist others, often in circumstances where it is not self-evidently to their advantage to do so. In the 1970s Rushton carried out a number of experiments on altruistic behavior, and he summarized these and integrated the empirical and theoretical literature in his 1980 book *Altruism, Socialization and Society.*[1] Rushton's experimental work showed that altruistic behavior is present among three to five year old children who often help one another in their play. He found that children's altruistic behavior is influenced by example in that they copy adults who behave altruistically, for instance by giving to others.

Rushton also found that altruism varies with sex and age, being typically stronger in women than in men, and increasing with age.

Theoretically, Rushton viewed the development of altruism in the framework of social learning theory. According to this approach, altruism is learned by children through the socialization practices adopted by their parents and by the influence of role models such as teachers and movie stars. Parents normally encourage and reward their children for behaving altruistically, while they discourage their children from behaving selfishly. These socialization practices foster the development of altruistic behavior. Children also copy the behavior of their parents and other adults through the process known as modeling. Rushton made some proposals about how altruism could be increased, for instance by improving moral education in schools and by providing more altruistic role models on television.

In the early 1980s Rushton became dissatisfied with the social learning theory approach to altruism. In the first place, he noted that individual differences were consistent and appeared early in life and led children to differ in their responses to socialization pressures and to role models. In 1983 he investigated the issue of a possible genetic contribution to individual differences in altruism during the course of a sabbatical year spent at the Institute of Psychiatry with Hans Eysenck. He used the London twin sample to assess the similarities between identical and non-identical twins for altruism and also on the related personality traits of nurturance, empathy, aggressiveness, and assertiveness. The results showed that identical twins were significantly more similar for all these traits than non-identicals, and that all the traits had heritabilities of between 50 and 60 percent.[2] Furthermore, the environmental factors affecting altruism appeared to be not the parental socialization techniques posited by social learning theory, but influences unique to each twin or what are technically called non-shared environmental factors. It was evident from this study that social learning theory provided a much less persuasive explanation for the development of altruistic behavior than Rushton had hitherto supposed. This result gave an impetus to Rushton's shift toward the sociobiological paradigm for human behavior.

GENETIC SIMILARITY THEORY

From around 1980 Rushton began to ponder the implications of sociobiological theory for human altruism. He thought that "kin-selection theory," so useful for explaining altruism in non-human animals, did not go far enough in the human case because humans also formed strong bonds outside the family, with intense friendships and group loyalties sometimes extending over lifetimes. Often these loyalties involve much self-sacrifice. Also, humans frequently behaved unaltruistically, to the extent of killing one another both on an individual basis and on a mass scale in warfare. As he was to note in 1984:

War has often directly and substantially affected
the gene pool, as when genocide was practiced (a
not uncommon occurrence during the long history of
Homo sapiens).[3]

Rushton's solution to these problems was his
Genetic Similarity Theory, published in 1984.[4]

Rushton's Genetic Similarity Theory states
that people tend to be altruistic toward those who
are like themselves and less altruistic or positively
hostile towards those who are unlike themselves.
Rushton began to formulate his Genetic Similarity
Theory in 1981 during a sabbatical year spent at
Berkeley, California. He noticed that in this multi-
racial community ethnic groups congregated
together and were concerned about their own kind.
Different ethnic groups typically sat with each other
in the college refectory and their college papers
were concerned with the problems of their fellow
ethnics, often in distant parts of the world. Black
newspapers were concerned with the plight of
blacks in England and in South Africa, thousands of
miles away. Jewish newspapers were concerned
with Israel and the problems of Jewish dissidents in
the former USSR. Hispanics lobbied to get Spanish
adopted as an official language and to increase the
numbers of immigrants from Mexico and Central
America. It was obvious that one of the most
important influences determining which issues
were regarded as important and what positions
would be taken on them was a person's own group
membership.

According to the Genetic Similarity Theory, individuals have inherited a tendency to like, seek out, and form mutually supportive friendships, marriages, and group memberships with those who are genetically similar because in the evolutionary past this helped to propagate copies of their genes more effectively. In a series of studies published in the late 1980s, Rushton showed empirically that people tend to choose spouses and friends on the basis of genetic resemblance measured by blood tests and the more heritable components of various attributes.[5] He summarized this work in 1989.[6]

Rushton's theory proposes four ways by which people detect others who are genetically similar to themselves. These are, first, "innate feature detectors," in which individuals inherit preferences for various attributes in others; secondly, "phenotypic matching," in which people inherit a tendency to learn to prefer others who share their attributes; third, through social interaction and familiarity; and fourth, through similarity of location. In these four ways, preference is typically given to the facial features, hair and skin color, language, accent, manners, attitudes, and other behaviors similar to an individual's own. These cues lead people to form mutually supportive in-groups sharing many of the same genes and to display indifference or hostility toward other groups, including other social classes, nationalities, and races with whom they have fewer genes in common.

Rushton's Genetic Similarity Theory is a new version of the theory of ethnocentrism, first proposed in the late 19th century by Herbert Spencer and kept alive in the 1960s by James Gregor, whose work is summarized in Chapter 9. Rushton has integrated the theory of ethnocentrism with more recent theories of kin selection and inclusive fitness, which state that individuals are biologically programmed to behave altruistically towards others to whom they are genetically similar, because this enhances the survival of the genes they have in common. Rushton's Genetic Similarity Theory is a synthesis of the classical ethnocentrism theory and modern sociobiology.

RACE DIFFERENCES IN INTELLIGENCE AND BRAIN SIZE

Rushton gave little thought to race differences in intelligence in the late 1960s while he was a student in London. He accepted the view of many social scientists that race differences are solely the result of environmental factors. He began to think about this issue in 1971, when Hans Eysenck published *Race, Intelligence and Education*. Eysenck endorsed the conclusions of Audrey Shuey and Arthur Jensen to the effect that the difference in average IQ between blacks and whites in the United States is significantly determined by genetic factors. Rushton read Eysenck's book and the research literature and concluded that the genetic case could not be ruled out but that the issue was unresolved.

Later in the 1970s Rushton thought more about the race-IQ question when he read Richard Lynn's work which showed that the Japanese in Japan had an average IQ of 107, seven points higher than that of American whites. He saw the force of Lynn's argument that the high IQ of the Japanese could not be easily explained by test bias or environmental advantage and that a genetic explanation was likely. He was also struck by Lynn's 1978 paper reviewing the world literature on intelligence levels in different races which showed that Chinese and Japanese consistently obtained somewhat higher IQs than whites, while Africans in Britain, the Caribbean, and Africa, as well as in the United States, scored considerably lower.

From these studies Rushton concluded that the problem of race differences in intelligence should not be considered only as a question of the causes of the IQ differences between blacks and whites in the United States but as a global phenomenon in which East Asian peoples consistently score a little higher on intelligence than Caucasians, who in turn score higher than blacks. In the 1980s Rushton began to work actively on this thesis. In October 1998 Rushton carried out a study of a highly selected group of black students attending the University of Witwaterstrand and the Rand Afrikaans University in Johannesburg, South Africa.[7] He found that black students in first year university averaged an IQ equivalent of only 85 [on the Standard Progressive Matrices]. Assuming that these students are one standard deviation above the general population, then the IQ for the general African population is indeed only about 70.

Rushton's principal empirical contribution was to show that there are East Asian—Caucasian—African differences in average brain size which run parallel to those for intelligence.[8] He re-examined the data on race differences in brain size collected in the 19th century by the American anthropologist Samuel Morton. Morton had a collection of skulls, whose volume he measured and categorized by race. He concluded that European Caucasoids had the largest average brain volume, followed by Mongoloids, while Africans had the smallest. Morton's data and methods were dismissed by Stephen Jay Gould in his 1981 book, *The Mismeasure of Man*, in which he claimed that Morton had juggled the figures to bring Caucasians out on top. Gould himself calculated the average brain sizes of Morton's skulls as 87 cubic inches for Mongoloids and Caucasoids and 83 cubic inches for Negroids. Gould concluded that the greater four cubic inches of brain possessed by Mongoloids and Caucasoids, as compared with Negroids, is trivial. Rushton argued that this is obviously not the case.

Rushton has reviewed the literature on race differences in average brain size and made calculations from five hitherto unanalyzed sets of data. The first consisted of the cranial capacities for 24 international military samples of males collated in 1978 by the U.S. National Aeronautics and Space Administration. After adjusting for the effects of height, weight, and body-surface area, the mean for East Asians was 1460 cubic centimeters and for Europeans 1446 cubic centimeters.[9] Next, in a random sample of 6,325 U.S. Army personnel measured in 1988 for helmet size, he found that after adjusting for the effects of body size, sex, and military rank, Asians, whites, and blacks had average cranial capacities of 1416 c^3, 1380 c^3, and 1359 c^3, respectively.[10]

Rushton then re-analyzed a set of anthropometric data originally published by Melville Herskovits as evidence against race differences in cranial size and calculated that Caucasoids averaged a cranial size of 1421 c^3 and Negroids, 1295 c^3.[11] Rushton also had data from tens of thousands of people from around the world collated in 1990 by the International Labour Office in Geneva and found that after adjusting for the effects of body size and sex, samples from the Pacific Rim, Europe, and Africa averaged cranial capacities, respectively, of 1308 c^3, 1297 c^3, and 1241 c^3.[12] Finally, in collaboration with Travis Osborne, he analyzed the Georgia Twin Study data and, after correcting for body size and sex, found that whites averaged a cranial capacity of 1269 c^3 and blacks 1251 c^3.[13]

Rushton's work on race differences in brain size has been a major contribution to the problem of the causation of race differences in intelligence. It has been shown in numerous studies that brain size is positively associated with intelligence. The existence of differences in average brain size between the races provides a neurological basis for at least some of the differences in intelligence.

RACE DIFFERENCES IN *r-K* REPRODUCTIVE STRATEGIES

From the early 1980s Rushton began to formulate the theory that the three major races of the Mongoloids, (principally Chinese and Japanese), the European Caucasoids, and the Negroids (African) can be ranked on a gradient which includes not only intelligence but a large number of social behaviors, and that this gradient could be explained in terms of racial differences in so-called *r-K* reproductive strategies. In 1984 he applied to the Social Sciences and Humanities Research Council of Canada for a grant to work out these ideas. The Council had previously supported his work, but on this occasion, his application was turned down. He sent the application to the Pioneer Fund, which has supported him from this time onwards.

Rushton's theory of race differences in *r-K* strategies was first proposed in 1985.[14] It has been elaborated in a number of subsequent papers and is summarized in his 1995 book *Race, Evolution, and Behavior*.[15] The theory starts with the biological concept that species differ in their reproductive strategies. One strategy *(r)* is to produce large numbers of offspring but devote little parental care to them; most of the offspring die young, but because there are so many of them enough reach maturity to assure species survival. The alternative strategy *(K)* is to produce few offspring but devote considerable parental care to rearing them, so that a much larger proportion survive. In general the first species to evolve (fish and reptiles) adopted *r*-strategies, whereas later species (mammals, especially primates) adopted *K*-strategies. For instance, frogs produce many hundreds of eggs, but female apes produce only one infant every 5 years or so. Those animal species that adopt the *K*-

strategy, especially monkeys, apes, and humans, have large brains and are more intelligent than r-strategists.

Rushton's innovative leap has been to apply r-K theory to the human races. His thesis is that humans evolved first in Africa as a moderately K-species. Some human groups subsequently migrated into Europe and Asia, where they evolved into the Caucasoids and Mongoloids. During this evolution they evolved stronger K-characteristics than the Negroid peoples who remained in Africa, and the Mongoloids evolved more pronounced K-characteristics than the Caucasoids. His theory proposes that the underlying Mongoloid—Caucasoid—Negroid gradient in r-K lifestyles explains their differences in brain size, intelligence, time of maturation, degree of sexual restraint, fertility, incidence of dizygotic twins, monogamy, mental health, lawabidingness, and anxiety.

Most of Rushton's work from 1985 onwards has been devoted to obtaining further evidence to test his theory of racial differences in r-K lifestyles. He has documented in exhaustive detail the presence of the predicted race differences in the United States, Canada, Britain, and throughout the world. In regard to crime, he has shown that in the United States, East Asians have low rates of crime, while blacks in both the United States and Britain commit crimes about five times more than whites. He has found the same pattern in international crime statistics, which show low crime rates in the Oriental nations of the Pacific rim, moderate rates in Caucasian countries and high crime rates among

African populations. Several analyses of crime data collected by Interpol, the International Criminal Police Organization, showed that violent crimes (murder, rape, and serious assault) were three times higher in African and Caribbean countries than in Pacific Rim countries and twice as high as in European countries.[16] For example, the international data for 1989-1990 showed that the average rates for murder, rape, and serious assault per 100,000 inhabitants were 143 for African countries; 74 for European countries; and 44 for Asian countries.

RACIAL DIFFERENCES IN SEXUAL BEHAVIOR

Rushton has extended his r-K theory to racial differences in a variety of sexual behaviors. He has documented evidence showing that Africans, African Americans, and blacks living in Britain are more sexually active, at an earlier age, and with more sexual partners than are Europeans and white Americans, who in turn are more sexually active, at an earlier age, and with more sexual partners than are Asians, Asian-Americans, and Asians living in Britain.[17] He has found that teenage fertility rates around the world showed the same racial gradient.

Rushton has also shown that the single-female-headed black family common in the Caribbean and the Americas, with father-absent households and lack of paternal certainty, is frequently present in sub-Saharan Africa. He argues that the widespread occurrence of this female-headed black family in many parts of the world shows that it cannot be attributed to the legacy of slavery, as often proposed by equalitarians. In a further discussion of these data, Rushton has suggested the African reproductive patterns may be partly determined by traits of temperament and biological factors such as the level of the sex hormone testosterone.[18]

Rushton wondered whether the race differences in sexual behavior and *r-K* lifestyles might show up in the incidence of AIDS in different populations. AIDS is contracted and spreads largely by sexual promiscuity, so it should be low among Mongoloid peoples, intermediate among Caucasoids, and high among Negroids. Rushton found that this is the case. In an examination of the 100,410 cases of AIDS reported to the World Health Organization as of 1 July 1988, African and Caribbean countries had a much greater number of AIDS cases per capita than those in Europe or Asia.[19] By 1 April 1990, the figures showed an 18 month doubling time, and by 1 January 1996, the figures had reached a cumulative total of 1,291,810 cases reported from 193 countries. Approximately 17 million people are estimated to have the human immunodeficiency virus (HIV) which causes the disease. Forty-seven countries are estimated to have 1 percent or more of their sexually active population living with HIV. Thirty-seven of these countries are in Africa, and seven are in the Caribbean. Within the United States, thirty million African-Americans have rates similar to their counterparts in black Africa and the black Caribbean with 3 percent of black men and 1 percent of black women with HIV. Among oriental

Americans, AIDS and HIV are negligible; white Americans fall intermediate blacks and Asians.

Rushton has also extended the theory of r-K race differences to the incidence of the birth of dizygotic twins. He shows that dizygotic twins are born most frequently among Negroids, with intermediate frequency among Caucasoids and with least frequency among Mongoloids. These differences in twinning rates are the most direct evidence for the basic proposition of the r-K race differences theory, to the effect that the human races are differentially biologically programmed in their propensities to have larger (Negroid), intermediate (Caucasoid) or smaller (Mongoloid) numbers of children. It can be argued that a number of the behaviors explained by the theory are determined by cultural factors, such as the different rates of crime, one parent families, sexual promiscuity, and the like, but it is much less easy to argue that cultural conditioning determines the birth incidence of dizygotic twins.

Rushton's imaginative and far-ranging theory extends the problem of human race differences well beyond the traditional concern of differences in intelligence. Rushton's theory posits race differences in intelligence as only one component of a much wider array of behaviors which have evolved differentially in the three major races of mankind, and provides an explanation for a number of racial differences, such as those in crime and marital cohesion, which are difficult to explain in terms of differences in intelligence alone.

Rushton's theory has attracted considerable interest, and some scholars have extended the theory to new data sets. Lee Ellis has argued that r-K theory explains why criminals tend to have a large number of siblings and half siblings, to come from broken homes, to have had shorter gestation times (more premature births), to have more rapid sexual development, a greater number of sexual partners, less stable marriages, to abandon, neglect, and abuse their children, and to have a shorter life expectancy.[20] Similarly, Richard Lynn has adduced data supporting the proposition that some forms of cancer are related to the racial r-K gradient. He notes that testosterone levels are a determinant of cancer of the prostate and that the incidence of this form of cancer should, therefore, be low among Mongoloids, intermediate among Caucasoids and high among Negroids. He has shown that this is the case.[21]

These proposals are amplifications of Rushton's theory, the wide scope of which can be likened to a jigsaw puzzle which is at present about half completed. The main outline of the picture can be seen but there remain a number of pieces still to be slotted in, and the completion of the picture will take a number of years.

RUSHTON COMES UNDER FIRE

It was inevitable that Rushton's theory of racial differences in r-K lifestyles would provoke hostility. The first wave of attacks occurred in January 1989 when Rushton presented his theory at the American Association for the Advancement of

Science meeting in San Francisco. The meeting was well attended by the press, who ran the story of Rushton's theory of race differences in a number of papers in the United States and Canada. The Canadian press gave the story the most attention.

The *Toronto Star* suggested in its 19 and 21 February 1989 editions that Rushton was a fraud, inept, or both, labeled the Pioneer Fund as Nazi, condemned as racist *The Mankind Quarterly* in which Rushton had published an article on altruism, and began a campaign to have Rushton dismissed from his position at the university. On 9 March 1989, *The Toronto Star* carried an editorial about Rushton entitled "A Weak Reaction to Academic Fraud," published a cartoon depicting him in a Ku Klux Klan hood and again demanded his dismissal. Shortly thereafter, in that paper's Easter Sunday editorial on 26 March 1989, the Holocaust and the Antichrist were linked to "an academic at an Ontario University." Rushton responded by taking libel proceedings against *The Toronto Star*. This brought the media campaign against him to a halt, whereupon Rushton dropped the legal action and took his case to the Ontario Press Council.

The legal actions against Rushton continued when 18 students lodged a formal complaint against him to the Ontario Human Rights Commission claiming that he had violated the . Human Rights Code guaranteeing equality of treatment to all citizens of the province. They charged him with "infecting the learning environment with academic racism." As remedy, the complainants requested that his employment at the university be terminated and that an order be made requiring the university "to examine its curriculum so as to eliminate academic racism." Four years after the complaint was lodged, the Ontario Human Rights Commission abandoned its case against Rushton claiming it could no longer find the complainants to testify. In an unrelated incident, the law was invoked to enable Canada Customs to seize copies of his book for 9 months while they determined whether to ban it as "hate literature."

The campaign against Rushton was taken up by the Ontario attorney general, who ordered a police investigation of his work for possible violation of Canadian "anti-hate" laws. A special combined Ontario Provincial Police and Metropolitan Toronto Police force concerned with pornography and hate literature set out to interview Rushton, along with senior members of the university administration and scholars in his department, to determine whether Rushton had violated the federal criminal code of Canada. After several months, on 3 November 1989, at a widely attended press conference, the attorney general issued his conclusion that Rushton's theories were "loony, but not criminal." During these travails, Rushton has received little support from the administrators of the University of Western Ontario.

Although publicly stating that the principle of academic freedom protected Rushton and his work, behind the scenes the administration made attempts to silence him and to remove his tenure. Rushton was given a negative annual performance rating, deprived of pay increases, investigated as to the ethics of his research work, convicted of minor breaches of protocol, and forced to deliver his lectures by videotape. He spent a year-and-a-half fighting the administration through a series of internal grievance procedures, eventually winning on the important issue of academic freedom.

Rushton has received a number of honors. He has been elected a Fellow of the American Association for the Advancement of Science, and of the American, British, and Canadian Psychological Associations. For the academic year 1987-1988 he was awarded a distinguished research professorship by his university with relief from teaching duties. For 1988-1989 he was elected a Fellow of the John Simon Guggenheim Memorial Foundation. In 1992 he was awarded a D.Sc. from the University of London. In addition to his work on sociobiology, Rushton has written, in collaboration with three colleagues, an introductory psychology textbook.[22] He has also written on the contributions of intelligence and personality to scientific productivity and creativity.[23] Despite the extensive harassment and attacks to which he has been subjected, Rushton remains resilient and continues to pursue his research on the genetics and sociobiology of race differences.

NOTES

1. Rushton, J. P. 1980. *Altruism, Socialization and Society.* Englewood Cliffs, NJ: Prentice-Hall.

2. Rushton, J. P., Fulker, D. W., Neale, M. C., Nias, D. K. B. & Eysenck, H. J. 1986. Altruism and aggression: The heritability of individual differences. *Journal of Personality and Social Psychology.* 50: 1192-1198.

3. Rushton, J. P. 1984. Sociobiology: Toward a theory of individual and groups differences in personality and social behavior. In J. R. Royce & L. P. Mos. (eds.) *Annals of Theoretical Psychology.* 2: 1-48. New York: Plenum Press. 9.

4. Rushton, J. P., Russell, R. J. H. & Wells, P. A. 1984. Genetic similarity theory: Beyond kin selection. *Behavior Genetics.* 14: 179-193.

5. Rushton, J. P. 1988. Genetic similarity, mate choice, and fecundity in humans. *Ethology and Sociobiology.* 9: 329-333; Rushton, J. P. & Nicholson, I. R. 1988. Genetic similarity theory, intelligence, and human mate choice. *Ethology and Sociobiology.* 9: 45-58; Rushton, J. P. 1989. Genetic similarity in male friendships. *Ethology and Sociobiology.* 10: 361-373.

6. Rushton, J. P. 1989. Genetic similarity, human altruism, and groups selection. *Behavioral and Brain Sciences.* 12: 503-559.

7. Rushton, J. P. & Skuy, M. 2000. Performance on Raven's Matrices by African and White university students in South Africa. *Intelligence.* 28: 251-265.

8. Rushton, J. P. 1988. Race differences in behavior: A review and evolutionary analysis. *Personality and Individual Differences.* 9: 1009-1024.

9. Rushton, J. P. 1991. Mongoloid-Caucasoid differences in brain size from military samples. *Intelligence.* 15: 351-359.

10. Rushton, J. P. 1992. Cranial capacity related to sex, rank, and race in a stratified random sample of 6,325 U.S. military personnel. *Intelligence.* 16: 401-413.

11. Rushton, J. P. 1993. Corrections to a paper on race and sex differences in brain size and intelligence. *Personality and Individual Differences.* 15: 229-231.

12. Rushton, J. P. 1994. Sex and race differences in cranial capacity from International Labor Office data. *Intelligence.* 19: 281-294.

13. Rushton, J. P. & Osborne, R. T. 1995. Genetic and environmental contributions to cranial capacity estimated in black and white adolescents. *Intelligence.* 20: 1-13.

14. Rushton, J. P. 1985. Differential *r-K* Theory: The sociobiology of individual and groups differences. *Personality and Individual Differences.* 6: 441-452.

15. Rushton, J. P. 1995. *Race, evolution, and behavior: A life history perspective.* New Brunswick, NJ: Transaction; (3rd edition). 2000. Port Huron, MI: Charles Darwin Research Institute.

16. Rushton, J. P. 1990. Race and crime: A reply to Roberts and Gabor. *Canadian Journal of Criminology.* 32: 315-334; Rushton, J. P. 1995. Race and crime: International data for 1989-1990. *Psychological Reports.* 307-312.

17. Rushton, J. P. and Bogaert, A. F. 1987. Race differences in sexual behavior: Testing an evolutionary hypothesis. *Journal of Research in Personality.* 21: 529-551.

18. Rushton, J. P. 1989. Race differences in sexuality and their correlates: Another look and physiological models. *Journal of Research in Personality.* 23: 35-54.

19. Rushton, J. P. & Bogaert, A. F. 1989. Population differences in susceptibility to AIDS: An evolutionary analysis. *Social Science and Medicine.* 28: 1211-1220.

20. Ellis, L. 1987. Criminal behavior and *r-K* selection: An extension of gene-based evolutionary theory. *Deviant Behavior.* 8: 149-176.

21. Lynn, R. 1990. Testosterone and gonadotrophin levels and *r-K* reproductive strategies. *Psychological Reports.* 67: 1203-1206.

22. Roediger, H. L. III, Rushton, J. P., Capaldi, E. D. & Paris, S. G. 1984. *Psychology.* Boston, MA: Little, Brown.

23. Jackson, D. N. & Rushton, J. P. (eds.) 1987. *Scientific Excellence: Origins and Assessment.* Beverly Hills, CA: Sage.

Robert A. Gordon, Ph.D.
Johns Hopkins University

Chapter 30

Robert A. Gordon

Robert A. Gordon (born 1932) is a sociologist who has devoted much of his career to the study of crime and its relation with social class and intelligence, and to the problem of why the crime rates of blacks in the United States are substantially higher than those of whites. He has also written on the issue of cultural bias in intelligence tests and on the relation of intelligence to the efficiency of job performance.

Robert A. Gordon was born in New York City in 1932. He attended the Bronx High School of Science and enrolled in the social science honors program at City College of New York. His studies were interrupted by a stint in the army, in which he served as a lieutenant in the infantry, after which he resumed his studies at City College, from which he graduated in the social sciences in 1957.

Gordon then entered graduate school in the sociology department of the University of Chicago as a Charles Richmond Henderson Fellow from 1957 to 1959, and a Louis E. Asher Fellow from 1959 to 1960. The Chicago social psychology program included faculty from both the sociology and psychology departments, and Gordon developed interests in social psychology, including small group behavior, which he found possessed a rich literature based on experimental work, field studies, surveys, and theory. Consequently, part of Gordon's graduate studies were devoted to psychology, taking courses in learning, human development in infancy and early childhood, the history of psychology, theories of personality, and personality measurement.

Gordon also studied attitude measurement, factor analysis, and multidimensional scaling in courses within the sociology department. Because he was aware of the close interdependence of psychology and sociology, he did not yield to the temptation to devote himself wholly to psychology, since his interest was in social problems, and especially in individual and group behavior as manifested in social deviance and the family; at that time such topics were being examined principally by sociologists. Although primarily a sociologist, Gordon's reading in psychology led him to the conclusion that sociologists were too resistant to relevant knowledge stemming from psychological research, and, in their concerns with grand theory, tended to be obstinately anti-reductionist. He determined to make every effort to meld the findings of the two disciplines in his own research. He joined both the American Psychological Association and the American Sociological Association.

While still a graduate student, Gordon became a research assistant and later associate director of a youth studies program on juvenile delinquency at the University of Chicago. This was a project studying the psychological and sociological characteristics of delinquent gangs and comparing these with control groups of non-delinquents. The project was staffed by more or less equal numbers of psychologists and sociologists. His daily involvement was mainly with the psychologists who had backgrounds in clinical, multivariate, and personality measurement psychology. Gordon also became familiar at this time with the semantic differential methods of Charles E. Osgood and his associates, and with Raymond Cattell's measures of personality and mental ability, which were used in the project. Gordon wrote both his master's and doctoral theses on this project. His master's thesis verified the factorial validity of Osgood's evaluation and potency factors for six different adolescent populations. His Ph.D. dissertation operationalized three theories of differences in values of gang delinquent and non-delinquent subcultures and tested these with a semantic differential instrument.

Gordon took up a position as an assistant professor in the sociology department at Johns Hopkins University, where for many years he taught courses in small group research and theory, juvenile delinquency, and methodology. He was appointed full professor in 1984. Most of his earlier publications at Johns Hopkins challenged sociology literature that erroneously denied the existence of an inverse relation between socioeconomic status and delinquency.[1]

STUDIES OF OPIATE ADDICTION

In the early 1970s Gordon became interested in opiate addiction, a topic which involved the interaction of biochemical factors with the sociology of deviant behavior. His work in this field, carried out in collaboration with a graduate student, W. E. McAuliffe, challenged the dominant symbolic interactionist theory of addiction that had excluded the possibility that positive reinforcement, that is, pleasurable effects, played a major role in continued drug use even after physical dependence had developed.[2]

Gordon's theory of addiction was based on the operant conditioning theory of B. F. Skinner. Gordon encountered considerable problems in getting this work published in the sociological journals. This was his first experience of the resistance to the publication of politically incorrect research conclusions all too common in the field of sociology. His theory of addiction conflicted with the accepted view which depicted addicts solely as victims of the withdrawal syndrome. Yet the evidence for the role of pleasurable effects accounted for aspects of addict behavior, such as relapse, drug preference for the more euphorogenic forms of opiate, and doses larger than necessary to avoid the withdrawal syndrome, that were not accounted for by the established theory, and held implications for the subsequent spread of recreational drug use.

INTELLIGENCE, DELINQUENCY, AND RACE

Gordon's interests in the connection between low intelligence and various social problems developed partly as a result of reading Arthur Jensen's 1969 *Harvard Educational Review* article and the intemperate reactions of many sociologists to Jensen's conclusions. In the early 1970s Gordon began to think about the role of low intelligence as a factor in crime, and of the possible relationship between the low average IQ of blacks and their high average crime rate. He constructed an index of crime which he called life time prevalence rate, consisting of the proportion of individuals born in the same year who qualify as delinquents by the age of 18 years. He then examined the research literature on crime rates and found that the life time prevalence rates for blacks was approximately three times as great as that for whites. For instance, in a study of appearances in the juvenile court in Philadelphia over the years 1949-1954, the life time prevalence for black males was 50.8 percent and for white males, 17.8 percent.[3] Gordon was unable to get papers on these conclusions published in mainstream sociology journals and had to get them published in the more statistical and mathematical publications.

Gordon noted that within both black and white populations intelligence is correlated with crime. He then calculated that the higher rate of crime among blacks could be entirely explicable in terms of their lower average IQs. He called this the "IQ-commensurability" phenomenon. Gordon argues that intelligence differences also explain the inverse relationship between socioeconomic status and crime, the higher socioeconomic classes having higher average IQs and lower crime rates. Mainstream sociological theory typically attributes crime to social class, as such, but Gordon has shown that IQ is a stronger correlate of crime than social status.[4]

CULTURAL BIAS IN IQ TESTS

Gordon's conclusion that low intelligence plays an important part in the high crime rates of the lower social classes and of blacks led him into the controversy about whether intelligence tests are biased against these two groups. His first work on this issue consisted of a critical examination of the "labeling theory" advanced by the sociologist Jane Mercer to explain the high proportion of black children in classes set aside for the educable mentally retarded in schools. Mercer's theory was that large numbers of these children scored low on intelligence tests because of the cultural bias of the test items. Gordon argued that the tests were not biased against blacks and that the large proportion of black educable mentally retarded children reflected a lower mean IQ of the black population.[5]

As a result of this work on cultural bias, Gordon was called as an expert witness for the defense in the 1978 case of *Larry P. et al. v. Wilson Riles, Superintendent of Public Instruction for the State of California, et al.* The plaintiffs in the cases argued that the higher frequency of blacks in special classes was due to cultural bias in the tests rather than to genuinely low intelligence of black children. The plaintiffs won the case, as a result of which the administration of intelligence tests to black children in California was outlawed.

Gordon continued to work on the test bias problem. There are several forms of the test bias theory, all of which require critical examination. One of these is that some items in the test are culturally biased against blacks. If this were so, then these items would be relatively more difficult for blacks and the rank order of difficulty of the items would be different for the two racial groups. Another way of expressing this is that there would be a group-by-item interaction effect. Gordon made a study of the research literature on this question going back to the results of the Stanford-Binet test given to the World War I army draft. He found that in the World War I data the item difficulties were virtually identical in blacks and whites, leading to the conclusion that test bias was not operating. This was also the case with more recent data.[6]

Gordon turned next to a consideration of the three backward digit span items among the 25 Stanford-Binet items. Digits items are unique because item difficulty can be increased without any fundamental change in item content and hence without introducing secondary facets affecting item behavior. His conclusion was that these items displayed virtually no item-group interaction for blacks and whites, a result that was extended to samples of juveniles from other IQ studies reporting digit span data for separate items.

Gordon argued that interaction-less items could be used as a probe for test bias by viewing them as anchors for consecutive points of difficulty only one digit apart, the minimum increment in difficulty allowed by their content. It was found that virtually the same Stanford-Binet items composed of far more heterogeneous content fell between consecutive digit items only one digit apart for both blacks and whites, even though the passing rates of all the items were much lower for blacks. Thus, the slightly greater amounts of item-group interaction in non-digit items were apparently a result of differences from item to item in secondary facets of content that were absent in the case of digits backward items.[7]

A second form of the test bias theory contends that blacks perform better in real life situations, such as in schools, colleges and attained occupational status than would be predicted from their IQs. If this were so, it would indicate that the tests had underestimated their potential and could be regarded as biased predictors. Gordon reviewed the literature of IQs and socioeconomic status among blacks and whites and found no differences. He found that regression lines of occupational status on IQ were parallel for blacks and whites, indicating that intelligence tests were equally valid predictors of occupational status for both races, contrary to the test bias theory.[8]

Gordon returned to the test bias problem in 1987 in a chapter in a book assessing the work of Arthur Jensen. Here he considered four variants of the test bias theory and found them all deficient. These were, first, the different item difficulty hypothesis; second, the hypothesis that IQ tests underpredict real life achievement of blacks; third, there is the hypothesis of "situational bias" which states that the testing situation impairs the performance of black children, either because the examiner is white, because blacks become over-anxious or for some other reasons; and fourth, there is the hypothesis that blacks perform worse on "culture bound" tests and better on "culture free" tests. Gordon argued that none of these variants of the culture bias theory stood up to examination.[9]

PREVENTIVE SENTENCING

In the late 1970s Gordon's interests in crime led him into a controversy over the policy of so called "preventive sentencing" of criminals. This involved the prediction of the potential dangerousness of criminals from their previous criminal record and taking this into account in determining the length of sentence imposed. Controversy had developed over preventive sentencing policies, particularly as they had been employed by Maryland's Patuxent Institution, to which Gordon was connected as member of the advisory board under a statute which required that the board include a member of his department. Patuxent, a maximum security institution to which dangerous repeat offenders were sent on indefinite sentences, had been caught up in the "de-institutionalization" movement of the 1970s, which was based on the view that estimates of the potential dangerousness of such criminals were often too severe, and that in consequence many individuals who would not prove dangerous in the future were being unfairly incarcerated.

The case against Patuxent contained a number of unexamined empirical, semantic, and philosophical issues, and Gordon now sought to evaluate these. On the empirical level, he called attention to the fact that Patuxent had been accused of overestimating the dangerousness of these repeat offenders, but comparison with studies of populations with much lower base rates of dangerousness showed that the critics of Patuxent had underestimated the dangerousness of the criminals incarcerated at Patuxent sometimes by as much as a factor of seven.[10]

Next, Gordon argued that if, as critics of Patuxent claimed, the success of predictions made at Patuxent in the cases of those criminals who were released was only about 50 percent, then it was almost equally appropriate to conclude that one could not predict safeness very well either. This implied that 50 percent of those who had been released should not have been released. This last aspect of the dispute, which had hitherto been ignored, was of considerable interest to the public and counterbalanced the argument that non-dangerous criminals were being wrongly held for crimes they might potentially commit. If one cannot determine which repeat offenders may be regarded as safe to let loose upon the public, and which were dangerous, then for practical purposes of decision-making one cannot effectively distinguish between them for they are all equally dangerous or equally safe. The question of interest then becomes, how dangerous are they on the average. Data supported the conclusion that Patuxent inmates were, on average, among the most dangerous individuals incarcerated anywhere in America, and were a thousand or so times more likely than the average white male to commit extremely vicious or harmful crimes. In the face of their prior criminal records it was unrealistic to judge Patuxent inmates on any theoretical "zero-one" dichotomy by which the probability of an individual's dangerousness was either zero or 100 percent. Gordon argued that the idea that any Patuxent inmate had a truly zero probability of proving dangerous was absurd.

IQ AND JOB PERFORMANCE

In 1988 Gordon took up the issue of the relationship between intelligence and the efficiency of job performance in a paper written jointly with Mary Lewis and Ann Quigley.[11] They began by noting the prejudice prevalent among sociologists against discussions of this issue, because of the widespread hostility in the discipline towards the concept of intelligence, to the acceptance of racial differences in intelligence, and to the capitalist system. Many sociologists, they note, typically attribute the social problems of blacks in the United States to the nature of capitalist society, to white racism, and to so-called "institutional racism."

The basic points about IQ, job performance, and race are that IQ is positively associated with the efficiency of job performance, and that because blacks obtain lower average IQs than whites, their job performance is less efficient. A number of critics of this view assert that the correlation between IQ and the efficiency of job performance is only around .3, and therefore intelligence explains only around 9 percent of the variance in job performance. Gordon and his colleagues point out that a roulette wheel gives the house a 2.7 percent advantage, equivalent to a .027 correlation in its favor, and yet this small advantage sustains a whole gambling industry. They conclude by asserting that intelligence is a significant predictor of job performance and that the low average IQs obtained by blacks explain why, on average, their job performance is less efficient than that of whites.

PROJECT FOR THE STUDY OF INTELLIGENCE AND SOCIETY

In 1986 Gordon set up the Project for the Study of Intelligence and Society at Johns Hopkins University, in collaboration with Linda Gottfredson, a graduate student (see Chapter 37). This project was financed by the Pioneer Fund. The objective was to carry out research on the contribution of racial differences in intelligence to crime and to employment, and to disseminate the publications on this research to professionals in the social sciences. In 1986 and 1987 Gordon and Gottfredson organized symposia on these issues under the auspices of the American Psychological Association.

Gordon and Gottfredson compiled a network of social scientists interested in these issues, and then they conducted several exploratory mailings.

Reprints of Gordon's paper, "Thunder from the Left,"[12] a review essay of Bernard Davis's *Storm Over Biology*,[13] and two issues of the *Journal of Vocational Behavior* edited by Linda Gottfredson were the first items mailed out. Soon the Project had over 1,000 academics and professionals from several different disciplines on its mailing list. It continues to inform social scientists with reprints of vital interest through the 1990s.

NOTES

1. Gordon, R. A. 1967. Issues in the ecological study of delinquency. *American Sociological Review*. 32: 927-944.

2. McAuliffe, W. E. and Gordon, R. A. 1980. Reinforcement and the combination-of-effects: Summary of a theory of opiate addiction. In D. J. Lettieri, M. Sayers, and H. W. Pearson (eds.) *Theories of Drug Abuse*: Selected Contemporary Perspectives. Washington, D.C.: Government Printing Office.

3. Gordon, R. A. 1973. An explicit estimation of the prevalence of commitment to a training school, to age 18, by race and by sex. *Journal of the American Statistical Association*. 68: 547-553.

4. Gordon, R. A. 1987. SES versus IQ in the race-IQ-delinquency model. *International Journal of Sociology and Social Policy*. 7: 92.

5. Gordon, R. A. 1975. Examining labelling theory: The case of mental retardation. In W. R. Grove (ed.) *The Labelling of Deviance*. New York, NY: Halstead-Sage.

6. Gordon, R. A. 1984. Digits backward and the Mercer-Kamin law: An empirical response to Mercer's treatment of internal validity of IQ tests. In C. R. Reynolds & R. T. Brown (eds.) *Perspectives on Bias in Mental Testing*. New York, NY: Plenum.

7. Gordon, R. A. & Rudert, E. E. 1979. Bad news concerning IQ tests. *Sociology of Education*. 52: 174-190.

8. *Ibid*.

9. Gordon, R. A. 1987. Jensen's contributions concerning test bias: a contextual view. In S. Modgil and C. Modgil (eds.) *Arthur Jensen: Consensus and Controversy*. New York, NY: Falmer Press.

10. Gordon, R. A. 1977. A critique of the evaluation of the Patuxent Institution, with particular reference to the issues of dangerousness and recidivism. *Bulletin of the American Academy of Psychiatry and Law*. 5: 210-255.

11. Gordon, R. A., Lewis, M. A. & Quigley, A. M. 1988. Can we muddle through the *g* crisis in employment? *Journal of Vocational Behavior*. 33: 424-451.

12. Gordon, R. A. 1988. Thunder from the left. *Academic Questions.* 1: 74-92.

13. Davis, B. 1986. *Storm Over Biology.* Buffalo, NY: Prometheus.

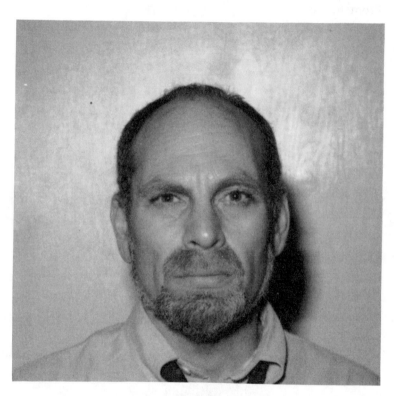

Michael Levin, Ph.D.
City College of New York

Chapter 31

Michael E. Levin

Michael Levin (born 1943) is the only professional philosopher to have been supported by the Pioneer Fund. As such, he has brought to the problems of race differences a mind trained in mathematical logic. He has also drawn on his expertise in moral philosophy to discuss the ethical issues involved in how race differences should be treated, and their effects moderated, in multi-racial societies.

Michael Eric Levin was born in New York in 1943. He attended Stuyvestant High School and Michigan State University where he majored in philosophy. He moved to Columbia University to do research on "Wittgenstein's Philosophy of Mathematics," for which he obtained his Ph.D. in 1969. As a post-graduate student he received several prestigious awards, including a Woodrow Wilson Fellowship, a Wilson Dissertation Fellowship, and a Columbia University Faculty Fellowship.

After receiving his Ph.D., Levin was appointed to the faculty of the philosophy department at the City College of the City University of New York. Levin's early career was devoted to the logical foundations of mathematics. He also wrote papers on the philosophy of science, metaphysics, epistemology, and metaethics. In 1979 he published *Metaphysics and the Mind Body Problem*.[1] In 1981 he received a National Endowment for the Humanities grant to pursue research on other problems.

In the 1970s Levin became interested in the social sciences and their application to social issues. He read widely and came to the conclusion that biological and genetic factors were more important determinants of individual and group differences than most social scientists and journalists were willing to admit. He began to articulate these conclusions in lectures, book reviews, and articles from the mid-1970s onwards. He took issue first with feminism and then turned to race differences, concluding that there are genetically based differences in intelligence and personality between men and women and between the races and that these are responsible for the differences in achievement and in crime.[2]

THE CRITIQUE OF RADICAL FEMINISM

Levin has criticized radical feminism in several articles, in his 1987 book *Feminism and Freedom* and in a paper written for the London based Institute of Economic Affairs.[3] Radical feminism asserts that there are no biologically determined psychological differences between males and females. Since men are everywhere overrepresented in positions of power and authority and have higher average earnings than women, feminists conclude that this is because girls are conditioned in childhood to be submissive and unambitious and because men operate policies of discrimination and sexism to keep women in a subordinate position. Feminists believe this is unjust and needs to be rectified by various policies of public intervention, such as by educating little girls to be more assertive and by affirmative action to give preference to women in appointments to top jobs.

Levin disputes all of these radical feminist assertions. He argues that men and women differ biologically and that biological factors are responsible for their different motivations and aptitudes for career success. He argues that men are innately more competitive, strongly motivated for career achievement and to make money as a symbol of success, whereas women are more strongly motivated to bring up children. Most women take time out from their careers to have children and rear them. This inevitably retards their progress up the career ladder to senior and highly paid positions. Women have children and spend time rearing them of their own free will, so the adverse effect of this on their careers and future earning capacity cannot be regarded as unfair. Levin writes that:

> if women earn less than men out of a clear-headed
> preference for other ends, there is no injustice ...
> women earn less because they prefer to stay at
> home.[4]

Conversely, men are more strongly motivated for
competitive success. They

> outrank women in the hierarchical world of work
> because they seek higher positions more avidly.[5]

Levin argues that the reason women
typically prefer to devote a considerable part of their
energies to bringing up their children rather than
single-mindedly advancing their careers is that
women are biologically inclined this way. He points
to a wealth of evidence showing that men are more
competitive and aggressive than women, and that
hormonal factors play an important role in this
difference.

In addition to these differences between men
and women in competitiveness, Levin also
proposes that there are differences in intelligence
that contribute to the higher earnings of men. He
notes that men are on average stronger on
mathematical and analytical abilities, which
command high incomes in economically
developed societies. Women's cognitive strengths
lie in clerical abilities and verbal facility which are
in less demand and hence less highly rewarded.
Levin argues that because the differences between
men and women in average earnings are

biologically determined they are not an injustice that needs to be put right by affirmative action.

Finally, Levin raises the question that feminism may have dysgenic effects. He argues that the feminist agenda of encouraging women to pursue careers rather than have children is responded to disproportionately by more intelligent women, who are remaining childless. He suggests that everyone should think seriously about this trend.

WHY RACE MATTERS

In 1991 the Pioneer Fund gave a grant to Levin to put together and develop further his ideas on race differences. Five years later Levin had completed a book, *Why Race Matters*.[6] The first part of the book sets out the evidence on race differences; the second discusses the evolution of race differences and the intense emotions aroused by the issue; and the third discusses the implications of the differences from the point of view of private and public morality.

Levin begins by setting out the evidence on race differences in intelligence. He argues that intelligence is accurately measured by intelligence tests, that there is a real and significant difference between blacks and whites in average IQs, that this cannot be wholly explained by environmental factors or by test bias, that the evidence points to a genetic basis for the difference, and that it is not crucial whether intelligence is conceptualized as a single entity called *g* or as a cluster of abilities. Most of this is a summary of the research literature, but

Levin makes an important new contribution in his analysis of the transracial adoption study carried out by Richard Weinberg and Sandra Scarr in which he shows that black babies adopted and reared in white families registered negligible gains in their IQs, contrary to the assertion of the authors of the study and the expectations of equalitarians and confirming the genetic basis of the differences.[7]

Levin makes another novel point regarding the differences between blacks and whites in athletic abilities. He points out the well known fact that American blacks do well in sports requiring fast reflexes and short bursts of energy such as sprinting, boxing, long jump, hurdling, and basketball, and he suggests that the explanation for this may be that blacks have more "fast-twitch" muscle tissue than whites. He notes further that blacks are not widely represented in sports requiring concentration and control such as tennis and golf, and that they are virtually absent from competitive games requiring high mental ability like chess, bridge, and Scrabble.

The usual theory advanced by equalitarians to explain the high level of achievement of blacks in certain sports is that these provide an escape out of impoverished environments. Levin argues that this cannot explain the distinctive type of sports in which blacks do well. He observes that equalitarians use the same argument to explain why Asians do well in math and science. But he asks why blacks select sports dependent on quick reflexes to climb out of the ghetto, while East Asians select math and science. Levin argues that the most plausible explanation for these differences is that blacks are naturally good at the sports requiring fast reactions, whereas East Asians are naturally strong on the abilities needed for achievement in math and science.

RACE DIFFERENCES IN PERSONALITY AND MOTIVATION

Levin turns next to an examination of race differences in personality, motivation, and temperament. He notes a number of striking differences between the races which are apparent to everyday observation. For instance, in music blacks tend to like jazz and rap with their strong rhythm and syncopation, while whites typically prefer music with a greater emphasis on harmony. He cites studies of personality showing that blacks score higher than whites on psychopathy, schizoid tendencies, and hyperactivity and points out that these traits are positively related to crime, suggesting that these traits may contribute to the higher crime rates and greater aggressiveness of blacks.

Levin examines the evidence on self-esteem among blacks. It has often been argued that blacks are relatively low on self-esteem and that some of their crime is motivated by a need to improve this. Yet he reviews research evidence showing that blacks tend to have higher self-esteem than whites. For instance, a much higher proportion of black than white girls are happy about their looks and figure. In questionnaire studies, blacks more frequently agree that "I am an important person," "I am actively self-confident," and similar statements. Levin agrees with J. P. Rushton that cross cultural studies show that blacks are less neurotic and more extraverted than whites and more permissive in their sexual attitudes and behavior, whereas east Asians deviate from whites in the opposite direction.

Levin's most original contribution in this area is his application of the economic theory of time preference to black-white personality differences. Economists have formulated the concept of time preference to describe the extent to which an individual is prepared to forego present advantages for future benefits. For example, some people are willing to save a significant proportion of their income, thereby foregoing the immediate pleasures that could be secured by spending it, in favor of the future advantages of having money in the bank, such as a more financially secure old age. Economists describe savers as having a low time preference, while those who prefer to spend their money here and now and are not as concerned about the future are described as having a high time preference.

Levin proposes that blacks tend to have higher time preferences than whites or, in other words, they are more "oriented towards the present" and have a stronger need for what in psychology is called "immediate gratification." He cites studies showing that when children are offered the choice between one candy bar today and two candy bars tomorrow, black children have a greater tendency to opt for one today.

Levin extends his idea that there are racial differences in time preferences to related characteristics. The high time preferences of blacks are associated with greater impulsivity and aggressiveness and lower levels of cooperation, rule following, and concern for the future. Levin generally opts for high impulsiveness as the most useful shorthand term to describe this syndrome of personality and temperamental characteristics prominent in blacks.

EVOLUTION OF RACE DIFFERENCES

Levin proposes an evolutionary theory to explain race differences in impulsiveness and associated characteristics. He begins by adopting Lynn's theory that the cold environments of Europe and Asia were more cognitively demanding than the tropical and semi-tropical environments of Africa, and that these greater cognitive demands produced the selection pressure responsible for the evolution of higher intelligence among the Caucasoid and Mongoloid peoples.

Levin proposes that the harsher environments of Eurasia would also have acted as a selection pressure for lower impulsiveness. As the Caucasoid and Mongoloid peoples became dependent on big-game hunting for their food supplies, they would have come under selection pressure to develop strong drives for group cooperation, rule-following, and the suppression of impulsiveness and aggressiveness towards other group members. In contrast, the Negroid peoples evolved predominantly as gatherers of plant foods, which were abundant in tropical and semi-tropical Africa and in which male group hunting activities were not necessary. For this reason there was less selection pressure on blacks to evolve strong controls over impulsiveness and associated characteristics. These different selection pressures led not only to differences in impulsiveness, but also to the evolution of different moral belief systems concerning the necessity of controlling impulsiveness. The Caucasian and Oriental peoples evolved moral systems in which the restraint of impulsiveness, aggression, rule-breaking, and dishonesty became strong moral imperatives. This occurred less among blacks and is a further reason, apart from intelligence, why black crime rates are higher than white.

Levin uses the philosopher Immanuel Kant's theory of moral values to explain further black-white differences in moral attitudes. Kant's principle of the golden rule states that people have a duty not to aggress or deceive others, and if necessary to help them. People are regarded as good to the extent that they conform to the ideals of the golden rule. Levin believes that the psychological evidence indicates that in terms of this principle the average black is not as good a person as the average white and that a greater proportion of blacks fall below the threshold of acceptable ethical behavior required by the golden rule.

Levin suggests that his evolutionary analysis helps us to understand why many whites are so sensitive about the research evidence indicating that blacks and whites differ genetically in regard to intelligence and impulsiveness. These whites regard intelligence and the control over impulsiveness as moral goods, so the evidence showing that blacks are weaker in these respects than whites is seen as deeply wounding to blacks. This offends the highly developed altruism of many whites and explains why they are so resistant to the research evidence on this issue.

In a further discussion of why the assertion that there are genetic differences between individuals and races evokes such violent emotional reactions, frequently consisting of accusations of fascism, Nazism, and the like, Levin suggests that we regard genetic characteristics as central to our personal identity in a way that environmentally caused characteristics are not. Thus, people would prefer to believe that they do not do well in school, in jobs, and so on because they have been handicapped through being brought up in a poor environment than because they are genetically unintelligent or lazy. The genetic explanation is resisted because it is more wounding to the individual's sense of self-worth.

POLICY IMPLICATIONS OF RACE DIFFERENCES

In the third part of his book Levin discusses the policy implications of the differences in intelligence and personality between blacks and whites. He deals with the problems of justice, crime, and the question of how far it is possible to treat people as individuals irrespective of their racial identity. With regard to justice, he discusses the moral basis of the various forms of affirmative action by which blacks are given preference over whites of equal or greater ability or qualifications in college admissions and job appointments.

Affirmative action policies, Levin argues, are based on a conception of justice that seeks to remedy past wrongs, in this case the wrongs perpetrated by whites on blacks, particularly those of slavery and of discrimination. He rejects this argument for affirmative action on the grounds that the whites who lose out when these policies are put into effect are not the ones who perpetrated the wrongs against blacks. The whites who perpetrated the wrongs were the ancestors of whites alive today, and it is impossible to justify punishing people for the wrongdoings of their ancestors.

Levin considers several other justifications for affirmative action. These are that although contemporary whites are not responsible for the past history of wrongs done to blacks, they do nevertheless secure advantages from these wrongs. He discusses the concept that groups as well as individuals can have rights; he considers the argument that if some blacks are appointed to prestigious and highly remunerated positions, they will serve as role models for young disadvantaged blacks who will benefit by having their aspirations raised. He rejects all these arguments for affirmative action.

Levin also questions the extent to which slavery in the United States actually harmed blacks. He argues that the proper comparison is how enslaved blacks and their descendants would have fared in Africa, if whites had not transported them to the United States. He points out that the transported blacks were already slaves in Africa, who were purchased from local chiefs by Europeans, and there is no reason to suppose that this was any better than being a slave in the United States. He points out that contemporary American blacks enjoy a much better education and a higher standard of living than blacks in Africa, and these are provided for them by whites. This makes it arguable that American blacks have benefited from their ancestors having been brought as slaves to the United States. Furthermore, many blacks continue to migrate voluntarily from Africa and the Caribbean into the United States, suggesting that they prefer life in America to that in their black countries of origin.

    Levin's conclusion is that American blacks do not experience any disadvantage that needs to be rectified by affirmative action. He believes that blacks have had equal opportunities to succeed in the United States for at least the last quarter of the 20th century and that their inability, as a population, to do so is a result of their lower average IQs and personality characteristics, not of discrimination against them by whites. He notes that in 1990 838 blacks and 35,199 whites obtained Ph.D.s, and he estimates that this is about the ratio that would be expected if an IQ of 130 is required to obtain a Ph.D. While there are some studies showing that blacks have lower achievement than whites even when their intelligence levels are the same, Levin believes that this can be explained by traits of impulsiveness and the desire for immediate gratification which impair educational achievement and job performance.

Levin's final point is that even if there is some merit in the claim that whites should compensate blacks for past injustices, whites have already compensated blacks sufficiently. He instances the Civil War, in which hundreds of thousands of Northern whites sacrificed their lives in order to secure the abolition of slavery. Furthermore, he argues that in contemporary times blacks are doing considerable damage to whites by crime and destruction of property. It is arguable, Levin concludes, that on these grounds blacks owe a collective debt to whites, rather than the other way round.

CRIME

Turning next to crime, Levin begins by documenting the evidence that in the United States the number of crimes committed by blacks is proportionately about 10 times as great as that committed by whites, and that approximately one in three young black males in their twenties has been incarcerated for a crime, as compared with approximately 2.5 percent of whites. In addition, about half of the assaults, murders, rapes, and robberies carried out by blacks are committed against whites. Levin concludes from these statistics that whites are rational in being frightened when they encounter young black males in situations where they might be attacked, such as when they are jogging in a park when there is no one else around.

Levin sees the moral problem as being what steps whites are justified in taking to lessen the dangers of being attacked by blacks. These consist of private actions taken by individuals and of collective actions taken by the police as agents of the state responsible for upholding the law. So far as private actions are concerned, Levin argues that whites are morally entitled to avoid blacks and to flee from them. For instance, a white cab driver is morally justified in refusing to pick up blacks because of the high risk of being attacked. It is true that this can be regarded as insulting to blacks, but Levin argues that this is a lesser evil than the cab driver's risk of being attacked.

Regarding actions carried out by the police, Levin argues that they are justified in being more suspicious of young black males than of young whites. For example, a policemen seeing a young black male driving an expensive automobile is entitled to stop and search him, because it is more likely that he is involved in drug dealing than is the case when an expensive automobile is being driven by a young white. Levin adopts the classical argument of Hobbes and Locke that one of the major functions of the state is to provide security for its citizens. The police know that blacks are much more likely to commit crimes than whites, and hence as part of their responsibility to protect citizens the police are obligated to investigate blacks for possible criminal activities more thoroughly than they investigate whites.

TREATING PEOPLE AS INDIVIDUALS

Levin turns next to the frequently made assertion that everyone should be treated as individuals, irrespective of their race. He argues that this apparently reasonable principle shrivels in the light of analysis. He notes first that this principle is violated by affirmative action which benefits blacks on the grounds of their membership with a race and does not treat them as individuals. He then argues that those who support affirmative action should also support police searches concentrated on blacks because in both cases individuals are being treated differentially on account of their race membership.

More generally, Levin argues that we are justified in judging others by reference to the categories they belong to because these are a guide to how they are likely to behave. For instance, he argues that a landlord would be justified in refusing to let accommodation to a member of a motorcycle gang if he had good reason to believe that gang members are likely to prove unsatisfactory tenants. Similarly, an employer may reasonably take the race of a job applicant into account if he has good reason to believe that members of some races are typically better employees than others.

In the final chapter of his book Levin argues that genetically based racial differences in intelligence and impulsiveness should be accepted rather than denied. He suggests that from a public policy point of view these differences are best dealt with according to the principles of a free market liberal society in which individuals are held responsible for their own lives with a minimum of support from welfare. He recognizes that this would bear more heavily on blacks because they are more dependent on welfare, but he regards this as the best practical solution to the problem.

TROUBLE AT CITY COLLEGE

Levin's views have generated considerable opposition at City College. In 1989 the dean of his faculty banned him from teaching some of his courses. In early 1990, the president of the college announced the formation of a committee to determine whether Levin's statements ranked as behavior sufficient to warrant the cancellation of his tenure. Levin's response to this was to obtain the assistance of the Center for Individual Rights, a public interest law firm, and sue his dean and the president of his college on First Amendment grounds, asking only that the committee be disbanded and his courses be restored, not for compensatory damages. A federal court sided with Levin. His college appealed, but Levin was upheld at the appellate level and was reinstated as a lecturer.

NOTES

1. Levin, M. 1979. *Metaphysics and the Mind-Body Problem*. New York: Oxford University Press.

2. Levin, M. 1990. Implications of race and sex differences for compensatory affirmative action and the concept of discrimination. *Journal of Social Political and Economic Studies*. 15: 175-212.

3. Levin, M. 1987. *Feminism and Freedom*. New Brunswick, NJ: Transaction; Levin, M. 1992. Women, work, biology and justice. In C. Quest (ed.) *Equal Opportunities: A Feminist Fallacy*. London: Institute of Economic Affairs.

4. *Ibid*. 13.

5. *Ibid*. 16.

6. Levin, M. 1998. *Why Race Matters*. Westport, CT: Greenwood.

7. Levin, M. 1994. Comment on the Minnesota transracial adoption study. *Intelligence*. 19: 13-20.

PART VI:
IQ, Population, and Social Policy

Seymour W. Itzkoff, Ph.D.
Smith College

Chapter 32

Seymour W. Itzkoff

Seymour Itzkoff (born 1928) has written a series of wide ranging books on the evolution of human intelligence, the emergence of racial differences and the significance of intelligence for the economic viability of modern nations. He has also issued warning about the decline of intelligence in western nations and proposed ways in which this deterioration could be reversed.

Seymour William Itzkoff was born in 1928 and raised in New York City. His parents were of Russian-Jewish origin, his father being a storekeeper, his mother working as a seamstress in the New York City garment industry. He attended New York public schools and entered the University of Hartford to study music and social science, playing the cello in the Hartford Symphony Orchestra while working for his bachelor's degree, which he obtained in 1950. After graduation, Itzkoff entered the U.S. Army and served as assistant principal cellist of the U.S. Army Symphony Orchestra in Washington, D.C. between 1950 and 1953. He was cellist in a piano trio that played at the Pentagon dinner held to celebrate the inauguration of NATO. Along with his wife Patricia, a violinist, he now performs with the Amicus Piano Trio.

Between 1956 and 1960, Itzkoff taught in the New York State public school system. At the same time he enrolled in graduate classes at Columbia University, where he received his master's degree in philosophy in 1956, and his doctorate in the same subject in 1964. After a brief period of teaching at Hunter College, he joined the faculty of Smith College in 1965, where he has spent the rest of his career.

Between 1969 and 1979, Itzkoff published five books in educational theory, philosophy, and musical biography. These included two books on the scientific philosophy of the neo-Kantian philosopher Ernst Cassirer which discussed a number of evolutionary theories with particular reference to issues concerning the significance of symbolic thought and knowledge in human societies.[1] This was the beginning of Itzkoff's interests in human evolution.

In 1969-1970 Itzkoff read the writings of Arthur Jensen and Hans Eysenck on the social importance of intelligence, its high heritability, and the probable genetic component in race differences. Concluding that they were correct, Itzkoff began to think about how race differences had evolved and the significance of intelligence and race for the contemporary world. With the assistance of grants from the Pioneer Fund from 1985 onwards he published seven books on these problems over the period 1983-1994.[2]

THE EVOLUTION OF HUMAN RACES

During the last quarter of the 20th century there have been two theories of the evolution of human races, known as the single origin theory and the multiregional theory. The single origin theory (now often referred to as the "Out of Africa" theory) holds that modern humans evolved in Africa around 140,000 years ago. Some groups from this ancestral population then migrated into Europe and Asia. They evolved into Caucasoids and Mongoloids. Subsequently further groups migrated into the Americas and into Australasia, where they evolved into the native American Indians and the Australian Aborigines and other southeast Asian peoples. This theory is favored by J. P. Rushton and a number of contemporary anthropologists.

The alternative multiregional theory proposes that pre-human *Homo erectus* populations migrated out of Africa and into Eurasia much earlier, around 1.5 million to 1 million years ago. *Homo erectus* had a brain size of about 1,000 to 1,300 c^3, considerably smaller than modern *Homo sapiens*. Their remains have been found throughout Eurasia. According to the multiregional theory, these *Homo erectus* populations in various parts of the world evolved independently into modern *Homo sapiens*. Itzkoff favors the multiregional theory, with modifications. He cites in support the continuity of certain morphological characteristics from *Homo erectus* to *Homo sapiens* peoples in different regions, for instance the 17 morphological characteristics described by Franz Weidenreich and exemplified in the shovel-shaped incisor teeth of *Homo erectus* specimens in northeast Asia which are also found in contemporary Mongoloids.

THE CRO-MAGNONS

Itzkoff proposes that a quantum leap in the evolution of human intelligence occurred in the Upper Paleolithic European Caucasoid or Cro-Magnon peoples, between 35,000 and 12,000 years ago. These appeared in Europe at the beginning of this period. Hitherto, Europe had been inhabited by the Neanderthals, now considered a distinct human species or subspecies. The Cro-Magnons migrated into Europe from the Near East and displaced the Neanderthals, either by killing them or driving them into remote and inhospitable regions where they were unable to survive.

The European Caucasoids displayed a higher level of intelligence than any previous humans and for this reason are designated *Homo sapiens sapiens*. The high intelligence of these peoples can be inferred from their technological and artistic achievements. They developed delicate double edged flint cutting tools with finely honed blades which were greatly superior to previous crude implements, and they invented the bow and arrow and the spear thrower. They invented the tent and were the first peoples to construct buildings with walls and roofs. In addition they showed great artistic ability in the form of the cave paintings which survive at Lascaux in southwest France and at Altamira in Spain.

Some of these European Caucasoids migrated into other parts of the world, particularly Africa and Asia. Here they encountered the African and Mongoloid peoples and frequently interbred with the women. In this way their genes, including those for intelligence, passed into the Mongoloid and African populations and spread among them. Itzkoff argues that this process took place more extensively among the Asians than the Africans, since the average intelligence level of the first group is much higher than that of the second.

Nevertheless, European Caucasoids remained the most intelligent population up to modern times. This is the explanation for the great scientific, technological, literary, and artistic achievements of these peoples during the last two-and-a-half thousand years, and for how they were able to colonize the Americas and Australasia and replace the indigenous populations. These successes constituted *The Triumph of the Intelligent*,[3] the title of the second volume of Itzkoff's series.

THE DECLINE OF THE EUROPEAN CAUCASOIDS

Itzkoff believes that the intelligence of the European Caucasoids in Europe, North America, and Australia has declined relative to that of the Mongoloid peoples during the 20th century as a result of wars, dysgenic fertility, and interbreeding with less intelligent peoples. He devotes his most recent book *The Decline of Intelligence in America* to answering this question.[4] He cites the decline in Scholastic Aptitude Test (SAT) scores, the growth of public debt and increasing crime and poverty in support of his view that average intelligence levels in the United States have declined. He asserts that the reason for this is that the poorest and least capable sections of the population have had more children than the more intelligent. He estimates that in the United States the mean IQ has dropped by about five points over the last several generations.

Itzkoff believes that dysgenic forces have been less powerful among the Japanese and Chinese in recent times and that this explains why in the 20th century these peoples obtain average IQs in the range of 103-107 in relation to the mean of 100 for European Caucasoids in Europe, North America, and Australia. Itzkoff maintains that the impressive economic achievements of Japan and the smaller Pacific rim economies of Taiwan, Hong Kong, South Korea, and Singapore are largely due to the high intelligence level of their populations.

THE NORTH-SOUTH DIVIDE

During the second half of the 20th century there has been much concern about the so-called north-south divide, that is, the disparity between the affluence of the economically developed and predominantly northern nations of the United States, Canada, Europe, and Japan, and the poverty of the predominantly southern third world, largely in South Asia, Africa, and much of Latin America.

The economically developed world has attempted to reduce this gap by giving economic aid and promoting education in the economically undeveloped countries, but Itzkoff argues that the results have been generally unimpressive. He believes that this is because the populations of the economically undeveloped countries are handicapped by their lower average intelligence.

Itzkoff argues this explains why Third World peoples fail to do as well as East Asians when they emigrate to the economically developed countries. Further, he asserts that much of the economic assistance given to Third World countries has been wasted because of the incompetence and corruption of the leadership.

EUGENICS

In his last two books, *The Road to Equality: Evolution and Social Reality* and *The Decline of Intelligence in America: A Strategy for National Renewal*,[5] Itzkoff examines the social policies responsible for the decline of intelligence in the United States and considers possible solutions. He attributes the decline largely to welfare programs which have encouraged the proliferation of the least competent, to liberal feminists who have encouraged intelligent women to pursue careers rather than have children, and to the large numbers of immigrants with lower average IQs.

To redress the decline of intelligence, Itzkoff argues that consciously designed eugenics measures are required. He proposes that the government should pass social policy legislation aimed at creating inducements, as well as legal protections, that will lead to the wealthy and successful having more than their share of children and the poor limiting their procreative activity in the interest of their own individual and social aspirations.[6]

Itzkoff's specific proposals are, first, that job priorities should be given to married men with families in order to encourage men to marry and have children. Second, all births should require the identification of the father, presumably in order to make fathers pay for their children and discourage them from producing children promiscuously. Third, high earning men and women without children should be penalized by a heavier burden of taxation. The objective of the government should be:

to establish a long-term social policy that will encourage the birth of 50 percent more children from the upper half of the social and income brackets than from the lower.

It is necessary, he writes in his most recent book:

to persuade the potentially parasitic classes at the top and at the bottom of society to act appropriately. The wealthy educated will have to validate their socially acquired assets by bearing their own offspring or adopting needy children. Those at the bottom should be humanely persuaded, with generous gifts if deemed

appropriate – but for one generation only – to
refrain from conceiving and having children.[7]

Itzkoff also believes that measures aimed at
discouraging illegitimate births in communities
would be useful. He supports the view put forward
by W. E. B. DuBois, the African American leader
and NAACP founder, who argued that the
"talented tenth" of the African-Americans should
have more children.

In addition, Itzkoff argues that the United
States should change its immigration policies. Only
the talented should be admitted as immigrants.
Illegal immigration should no longer be tolerated
or condoned by amnesties and "those who are here
in violation of our laws, along with the children
that have been born here in the interim," should be
repatriated.[8]

In the late 1980s, Itzkoff's forthright views
began to encounter considerable opposition at
Smith College. By 1990 this reached the point at
which the College proposed to refuse to accept
further grants from the Pioneer Fund. Itzkoff
threatened legal action against the College. In the
dispute that followed, the American Association of
University Professors supported Itzkoff, and in the
end the College agreed to accept such grants. In
addition, the College faculty reaffirmed the right of
academics to receive funds for research free from
administrative interference and veto. Fortunately,
the attack on academic freedom proved
unsuccessful.

In 1994, following a *New York Times Sunday Book Review* praising Itzkoff's *The Decline of Intelligence in America*, radical students rioted against him at Smith College. Itzkoff's office building and personal office library were entered, vandalized, and spray painted. His home was also attacked. The Smith College administration stood aside at the demonstrations against their "tainted" senior professor. Only belatedly did the College administration express its regrets at the violence. Police never apprehended the perpetrators. The Smith College faculty unanimously passed a resolution condemning these violations of academic freedom and the terror they inspired.

NOTES

1. Itzkoff, S. W. 1971. *Ernst Cassirer: Scientific Knowledge and the Concept of Man*. Notre Dame, IN: University of Notre Dame Press; Itzkoff, S. W. 1977. *Ernst Cassirer, Philosopher of Culture*. Boston: G. K. Hall.

2. Itzkoff, S. W. 1985. *The Triumph of the Intelligent: Creation of Homo Sapiens Sapiens*. New York: Peter Lang International Publishers; Itzkoff, S. W. 1987. *Why Humans Vary in Intelligence*. New York: Peter Lang International Publishers; Itzkoff, S. W. 1990. *The Making of the Civilized Mind*. New York: Peter Lang International Publishers; Itzkoff, S. W. 1991. *Human Intelligence and National Power: A Political Essay in Sociobiology*. New York: Peter Lang International Publishers; Itzkoff, S. W. 1992. *The Road to Equality: Evolution and Social Reality*. New York: Praeger; Itzkoff, S. W. 1994. *The Decline of Intelligence in America*. Westport, CT: Praeger.

3. Itzkoff, S. W. 1985. *Op. cit.*

4. Itzkoff, S. W. 1994. *Op. cit.*

5. Itzkoff, S. W. 1992. *Op. cit.*; Itzkoff, S. W. 1994. *Op. cit.*

6. *Ibid.*

7. Itzkoff, S. W. 1994. *Op. cit.* 192-195.

8. *Ibid.* 161.

Chapter 33

Daniel R. Vining

Daniel Vining (born 1944) is a demographer who has carried out research on the low fertility of those with better education, higher income, and higher IQ that has accompanied the demographic transition in the economically developed nations. He has documented the low fertility that appeared in the last quarter of the twentieth century throughout the economically developed world. He has also written on China's pro-eugenics policy.

Daniel R. Vining was born in 1944, in Fayetteville, where his father taught economics at the University of Arkansas. In 1945, he moved to Charlottesville, Virginia, where he grew up and where his father was a professor of economics at the University of Virginia. Vining went to Yale in 1961 and graduated in 1966 with a B.A. in philosophy. He spent the years 1963-1964 in the U.S. Marine Corps Reserve. In 1966, he went to Vietnam, where he stayed for two years, working for the

International Voluntary Service. In 1968, he went to Japan, where he taught English at the Tokyo English Center. In 1969, Vining entered the Woodrow Wilson School at Princeton, from which he graduated in 1971 with a master's in public affairs. He spent the summer of 1970 in India working for the Agency for International Development. In the fall of 1971 he went to Carnegie-Mellon University in Pittsburgh, where he obtained his Ph.D. in 1975 in urban and public affairs.

In 1974, Vining was appointed to the faculty of the University of Pennsylvania to teach statistics in the Regional Science Department. He worked at the Office of Population Research at Princeton in the year 1979-80, and took a sabbatical at the University of Glasgow in 1982. He has also worked in the Population Studies Center at the University of Pennsylvania.

Vining spent the years 1974-80 working on the study of spatial population concentration. When a country becomes developed, this concentration stabilizes or even reverses. In the developing state, the tendency is for a population to concentrate spatially. Vining showed this to be the case, with a grant from the National Science Foundation.

In the late 1970s Vining became interested in the demography of fertility and the question of whether there was an inverse relationship between fertility and intelligence. In 1981 he obtained a grant from the Pioneer Fund to investigate this issue, and he was supported by the Fund from that year through 1993.

DYSGENIC FERTILITY

Vining's work on the relationship between intelligence and fertility in the United States was the first major study on individuals to demonstrate that the more intelligent tend to have fewer children than the less intelligent, and therefore that fertility is dysgenic and the quality of the population is deteriorating.[1] Vining analyzed data for a sample of approximately ten thousand individuals born between 1942 and 1954 and calculated the correlations between their IQs and their number of children. He found the correlations were significantly negative for white women (-.18), black women (-0.20), and for white men (-0.14); the data were insufficient to calculate a correlation for black men.

The results of Vining's study agree with the greater dysgenic fertility for women than for men found in several other studies. A follow-up of the sample showed the persistence of dysgenic fertility in the sample when they were older and had completed their childbearing years.[2] Vining has also investigated the relationship between intelligence and fertility in Japan and in Sweden.[3]

THE SOCIOBIOLOGY OF DYSGENIC FERTILITY

Vining has discussed the presence of dysgenic fertility from a sociobiological perspective.[4] He argues that the fundamental postulate of sociobiology is that people strive to maximize their number of children. In most human societies high status men have acted in conformity with this postulate and had more children than low status men. However, in the 20th century this postulate has broken down in the economically developed nations; high status men, who also tend to be more intelligent, have had fewer children than low status men.

,According to Vining, Frederick Osborn's "eugenic hypothesis" that dysgenic fertility would be only a temporary phase of the demographic transition has turned out to be unfounded and that dysgenic fertility shows no sign of coming to an end. Vining suggests that intelligent women have fewer children than unintelligent women because they are more efficient users of contraception, because they tend to delay childbearing to a period of their lives when fecundity declines, and because of the competing demands of their careers. He notes that phenotypic intelligence has been increasing during much of the 20th century and that this must be due to environmental improvements acting in the opposite direction to dysgenic fertility. However, he thinks it unlikely that these environmental improvements can continue indefinitely and that when their input is exhausted, phenotypic intelligence will begin to decline.

BELOW REPLACEMENT FERTILITY

In addition to persistent dysgenic birth rates, Vining sees a second problem in contemporary demographic trends. This is that in the economically developed countries women are using modern methods of contraception so effectively that they are not having sufficient children to maintain the size of the population. This has led to the appearance of below replacement fertility in economically developed nations. Below replacement fertility means that the population as a whole has too few children to replace itself. Demographers estimate that to keep the population at a constant size the average woman has to produce approximately 2.1 children. It might be expected that the reproduction of 2.0 children by every woman would be sufficient to keep the population size stable, but an additional 0.1 is required because more boys are born than girls and because some women die before reaching the end of their childbearing years.

Vining believes that intelligence must have evolved in hominids over a period of several million years because it conferred a fitness advantage, that is to say the more intelligent left more surviving offspring than the less intelligent. However, this evolution contained the seeds of its own destruction because eventually humans developed the ability to control their own fertility, and in the late 20th century they have become so efficient at doing this that they are no longer replacing themselves.

Vining has published extensive data on the presence of below replacement fertility in North America, Europe, and East Asia from 1965 through 1987.[5] His figures show that in 1965 fertility was generally above replacement in all these regions. In the United States it was 2.29 and in Canada, 3.15; in

Western Europe it lay between 2.51 (West Germany) and 4.02 (Ireland), and in East Asia it was 2.08 in Japan and 2.52 in Singapore. Only in two countries of Eastern Europe was fertility below replacement, namely Hungary (1.82) and Romania (1.91). By 1980 fertility had fallen to below replacement throughout virtually the whole of the economically developed world, and it continued to fall until the late 1980s, when it stood at 1.84 in the United States and 1.67 in Canada. In Western Europe fertility had fallen well below replacement, except in Ireland, Poland, and Russia; and the same was true of Japan and Singapore. Vining concluded that the demographic transition from high to low fertility appears to lead to what he called a "demograph trap" in which the population fails to replace itself.

It is frequently argued that this problem can be overcome by immigration, so that immigrants will augment the indigenous stock of North America, Europe, and the Far East. One of those who favors this solution is Ben Wattenberg who welcomes the growing racial and ethnic diversity of the United States in his book *The First Universal Nation*.[6] Vining reviewed Wattenberg's book and saw difficulties in the immigration solution because of the endemic racial and ethnic conflicts which are present in all multi-ethnic and multi-racial societies.[7] He wonders why Wattenberg thinks that the United States will be immune from these conflicts which are so common elsewhere in the world. He has made a study of some of the problems of a multi-ethnic society by examining the case of Israel.[8] Here, as a result of its higher fertility, the Arab population is quickly outgrowing the Jewish population, and Vining estimates that by around the year 2070 Arabs will outnumber Jews in Israel.

EUGENICS IN CHINA AND SINGAPORE

While by 1970 eugenics had largely fallen out of favor in North America and Europe, it continued to be accepted in China and Singapore. Vining has followed the standing of eugenics in these two countries.

In 1984 Vining published an account and translation of the eugenic views of the Chinese demographer Sun Dong-Sheng which appeared in 1981.[9] Dong-Sheng begins by stating that:

> Naturally, if one wishes to see that every family is able to produce healthy, intelligent children, then it is necessary to study eugenics, to popularize the knowledge of this field and to master its principles.[10]

He devoted most of his paper to the prevention of genetic diseases and disorders. He noted that the birth incidence of these is higher among close relatives and stated that in China the children of first cousins have 150 times the incidence of genetic diseases than the children of unrelated couples. He stated that Chinese geneticists estimated that the prohibition of first cousin marriages would result in a 20 percent drop in the birth incidence of deaf mutes and a 15 percent drop in the birth incidence of amaurotic idiocy. In 1980 China introduced a Marriage Law prohibiting marriages between collateral relatives within three generations because of the high incidence of genetic disorders among the offspring.

The new law also prohibited the marriages of those with certain diseases including mental illness, tuberculosis, and serious heart, liver, and kidney defects. In cases where people with these diseases were already married, pregnant wives were required to undergo prenatal diagnosis and abortion of defective fetuses.

Dong-Sheng also recommends the more widespread use of prenatal diagnosis of women at risk of having a defective child, such as women over the age of 35 having a child with Down's syndrome, and abortion in cases where a defective fetus is diagnosed. He estimates that approximately 10 million children in China suffer from some kind of serious genetic defect and that:

> much parental anguish is caused by these children; they are unable to do anything useful; they are a financial and mental burden on their parents; and they pose an increasing burden on our country. Socialist modernization urgently needs a reduction or elimination of genetic disease and hereditary defects. Only by promoting the births of better offspring can we improve the genetic quality of our population, reduce or eliminate a variety of genetic diseases, and thereby lessen the burdens imposed on both family and nation. Therefore, to promote eugenics is to secure immeasurable advantages with no harmful consequences.[11]

He concludes that:

> It is our earnest hope that eugenics should not be construed as a purely expedient measure, but rather as a long term mission which concerns the long-term prosperity of the Chinese race for centuries ahead. Each one of us ... must actively study and propagate the knowledge of eugenics and bring about the birth of healthier, superior children.[12]

Another Chinese demographer who has expressed favorable views on eugenics and whose work Vining has summarized is Yuan Ryoyun.[13] In a paper published in 1983, Ryoyun affirmed that:

> The enhancement of the quality of China's population ... is the objective need and destiny of China's socialist population development.[14]

He limits his discussion of eugenics to measures designed to reduce the incidence of genetic diseases. Ryoyun proposes several measures to achieve this objective, including the greater use of prenatal diagnosis for genetic diseases and abortion of defective fetuses and premarital examinations for genetic disorders among those planning to get married and prohibitions of marriage and having children on those with a range of genetic disorders. He also argued that China should endeavor to improve the environmental conditions affecting the fetus and child, for instance by better obstetric care, preventing pregnant women from smoking or having excessive alcohol consumption, and encouraging them to improve their nutritional intake. He argued that "We must popularize and propagate eugenics" and "eugenic laws need to be formulated."[15] Ryoyun concluded by quoting a speech of Chinese premier Zhao Ziyang in which eugenic measures were urged for the improvement of the quality of the Chinese population.

Vining suffered a stroke in 1985, which disabled him physically and impaired his work productivity. He has shown fortitude in the face of this misfortune and has continued to publish articles on demography, dysgenics, and below-replacement fertility.

NOTES

1. Vining, D. R. 1982. On the possibility of the reemergence of a dysgenic trend with respect to intelligence in American fertility differentials. *Intelligence.* 6: 241-264.

2. Vining, D. R. 1995. On the possibility of the reemergence of a dysgenic trend: an update. *Personality and Individual Differences.* 19: 259-265.

3. Vining, D. R., Bygren, L., Hattori, K., Nystrom, S. & Tamura, S. 1988. IQ-fertility relationships in Japan and Sweden. *Personality and Individual Differences.* 9: 931.

4. Vining, D. R. 1986. Social versus reproductive success. *Behavioral and Brain Sciences.* 9: 167-216.

5. Vining, D. R. 1988. Below replacement fertility in five regions of the world. *Mankind Quarterly.* 29: 211-220.

6. Wattenberg, B. J. 1991. *The First Universal Nation.* New York, NY: Free Press.

7. Vining, D. R. 1994. Surveying multi-ethnicity. *Mankind Quarterly.* 35: 151-156.

8. Vining, D. R. 1989. The demographic problem in Israel. *Mankind Quarterly.* 30: 65-70.

9. Dong-Sheng, S. 1984. Eugenic thinking in contemporary mainland China. *Mankind Quarterly.* 24: 167-183.

10. *Ibid.* 170.

11. *Ibid.* 182.

12. *Ibid.* 183.

13. Desilets, M. & Vining, D. R. 1984. Quality considerations in population policy. *Population and Development Review.* 10: 753-761.

14. *Ibid.* 756.

15. *Ibid.* 780.

Robert D. Retherford, Ph.D.
East-West Population Institute

Chapter 34

Robert D. Retherford

Robert Retherford (born 1941) is a demographer who has worked on the decline of the intelligence of the populations of the United States and Japan caused by the greater fertility of people with lower IQs and income. He has formulated a general theory of the onset of dysgenic fertility during the demographic transition from large to small family size, and of its persistence in the last two decades of the twentieth century.

Robert Dennis Retherford was born in 1941 in New York City. He entered the University of California at Berkeley and in 1964 obtained his B.A. in physical sciences and pre-medical studies. He remained at Berkeley for his postgraduate work, which he undertook in sociology and earned a master's degree in 1966 and a Ph.D. in 1970. He held a Population Council fellowship during the years 1967-1968 and 1968-1969.

Retherford specialized in demography and worked at the Institut National d'Etudes Demographiques in France on a Social Science Research Council postdoctoral research training fellowship during the academic year 1970-1971. In 1971-1972 he served as a consultant to the population division of the United Nations Economic Commission for Asia and the Far East.

In 1970 Retherford joined the East-West Population Institute in Honolulu as a research associate. He was appointed assistant director of professional education in 1974 with responsibility for organizing conferences and internship programs, and assistant director for graduate study in 1980, supervising a scholarship program involving some forty graduate students at any one time. In addition to these duties, Retherford has carried out research on demographic trends in the United States and Asia. He has served as president of the Society for the Study of Social Biology and editor of the Asian and Pacific Census Forum. He is an affiliate member of the faculty at the University of Hawaii, where he teaches graduate courses in population studies.

Retherford's demographic research and activities in organizing international seminars have been funded by grants from the U.S. National Academy of Sciences, the Rockefeller Foundation, the Registrar-General of India, the U.S. Bureau of the Census, and the National Institute of Child Health and Human Development. He has also received grants from the Pioneer Fund in 1984 and 1994 to carry out research on the demographic transition, the evolution of intelligence, and dysgenic fertility for intelligence and educational level.

DYSGENIC FERTILITY IN THE UNITED STATES

In 1988 Retherford published one of the most thorough studies of the relationship between the intelligence of adults and their number of children in the United States.[1] This study, carried out in collaboration with William Sewell, was based on a sample of 10,317 high school seniors in Wisconsin for whom IQs had been obtained in 1957. They were followed up in 1975, and the number of children they had had was ascertained. Their number of children was then examined in relation to their intelligence levels. The result showed an inverse relationship between their IQs and their number of children; that is, the trend was for the more intelligent to have had fewer children. For instance, among the women, those in the lowest intelligence group with IQs in the range of 67-81 had an average of 2.76 children, while those in the intelligence group with IQs between 121-145 had an average of 2.29 children. Among men, those in the lowest IQ group had an average of 2.36 children, while those in the highest IQ group had an average of 2.07 children.

Because more intelligent people are having fewer children than less intelligent ones, these fertility differentials imply that the genetic quality of the population in respect of intelligence must be deteriorating. Retherford and Sewell calculated the magnitude of this deterioration. First, they estimated the difference between the average IQ of their sample and that of their sample's children. This was done by assuming that the children's intelligence is on average the same as that of their parents, and weighting the IQ of the parents by their number of children. The result was that the average IQ of the children was 0.81 IQ points lower than that of the parents. This difference is called the selection differential and is the decline in intelligence between the two generations which would be present if intelligence were solely determined by genetic factors. In order to estimate the genetic deterioration the selection differential is multiplied by the heritability of intelligence. Retherford and Sewell use a figure of 0.4 for this so the deterioration of genotypic intelligence becomes -0.32 IQ points per generation (-0.81 x 0.40 = -.32). Most authorities in this field would consider the heritability of 0.4 used by Retherford and Sewell too low and 0.8 as being about the right figure. Adoption of the higher heritability coefficient doubles the rate of genotypic deterioration to -0.64 IQ points per generation.

Retherford and Sewell also found that the dysgenic fertility is greater for women than for men. The selection differential for women was -1.33, while for men it was only -0.28. Probably the reasons for this are that in their early adult years many women postpone childbearing in order to pursue their careers. By the time they reach their thirties they find, for various reasons, that they have delayed having children too long. Men, on the other hand, find it easier to combine having children with their careers or alternatively find it relatively easy in their thirties and forties to find younger wives to have their children.

In a further paper based on this sample, Retherford and Sewell consider why intelligence is negatively associated with fertility. They conclude that the effect operates largely through education. More intelligent people generally obtain more education and enter professional careers, which leads them to postpone childbearing and to have relatively few children.[2]

FERTILITY IN JAPAN

Retherford has worked on fertility in Japan in collaboration with Japanese colleagues.[3] He has documented the decline in family size in Japan from 1950 to 1995, from approximately 3.7 to 1.5 children per woman. Survey data on ideal family size show that this has remained approximately constant at around 2.4 children from 1960. Thus women are having significantly fewer children than they consider ideal. The principal reasons for this seem to be the costs and stress of raising and educating children.

Retherford has also examined the issue of whether fertility has been dysgenic in Japan.[4] It has not proven possible to tackle this question directly by obtaining data on intelligence and numbers of children, but he has examined it indirectly from evidence on the relationship between educational level and fertility. This can be done because educational level is strongly associated with IQ, so it can be regarded as an indirect measure of intelligence.

Retherford and his Japanese colleague Naohiro Ogowa have analyzed survey results in which married women aged 40-49 were divided into those with primary, secondary, and tertiary education, and their average numbers of children calculated. The data were collected each year from 1963 through 1992. The results showed that those with primary education had more children and those with tertiary education had fewest in the survey years 1963-1981, indicating dysgenic fertility, but from 1984 to 1992 there were no significant fertility differentials. This appears to show that dysgenic fertility has ceased in Japan. However, the survey was conducted only on married women, and greater numbers of Japanese women with tertiary education do not marry and are childless. Hence the conclusion to be drawn from the study is that dysgenic fertility has slackened in Japan in the last two decades of the 20th century, but has not disappeared.

THE HISTORICAL DEVELOPMENT OF DYSGENIC FERTILITY

Retherford has formulated a general theory of the historical development of dysgenic fertility.[5] The principal components of the theory are that in previous historical times the higher socioeconomic classes had greater average fertility than the lower. The higher socioeconomic classes also tended to be more intelligent, because intelligence contributes to upward social mobility. The effect of this was to make fertility eugenic. With the onset of the demographic transition in the economically developed world in the 19th century, there was a general reduction of fertility. This reduction was initially greatest in the higher socioeconomic classes. This was dysgenic and entailed a reduction in genotypic intelligence (the genetic quality of the population).

The theory proposes five principal reasons why the higher socioeconomic classes reduced their numbers of children earlier and to a greater extent than the lower. First, the higher socioeconomic classes generally had lower child mortality, so they had more surviving children than they wanted. Second, children were more costly for them, because the higher social classes were expected to educate their children more extensively, whereas the lower socioeconomic classes' children could be put to work at a young age and were an economic asset. Third, the higher socioeconomic classes had more competing interests with which children interfered and so had stronger motives to curtail their numbers of children. Fourth, opinion leaders in favor of small families had more influence on the higher socioeconomic classes. And fifth, the higher social classes were the first to use modern methods of contraception.

Retherford notes that as the demographic transition has progressed, socioeconomic differences in fertility have narrowed. This has reduced the dysgenic effect, but it has not disappeared. He proposes a number of reasons for the narrowing of socioeconomic fertility differences: the deliberate curtailment of fertility has spread through the whole of society; child mortality has become very low in all social classes; children are no longer permitted to work, so they have ceased to be an economic asset for the lower socioeconomic classes; and the value of children as a means of support in old age has diminished. In the late 20th century the majority of the populations of economically developed nations have come to regard having two children as the ideal, and many of them achieve this objective.

Nevertheless, some dysgenic trend is still present in all countries in the last two decades of the 20th century. Retherford believes that the principal reason for this lies in the tendency for well educated women to pursue their careers rather than have children. Because well educated women have high IQs, the effect of this is dysgenic. Retherford's paper on dysgenic fertility in modern times and of the resulting deterioration of genotypic intelligence was delivered in 1993 at a conference held by the Society for the Study of Social Biology (formerly, the American Eugenic Society). At the time of this writing the paper has not been published.

Retherford's most recent work, carried out in collaboration with Norman Y. Luther, considers the question of whether fertility differentials by education have been converging in the United States. According to his theory of the demographic transition, fertility differentials by education tend to become strongly negative in the early stages of transition, because family limitations tends to be practiced first among the more educated. As the transition proceeds, contraceptive use diffuses to the less educated, and fertility differentials by education tend to converge. However, the evidence from this study indicates that in the United States fertility differentials by education have not appreciably converged. As late as 1990, the latest year considered, fertility differentials by education were still strongly negative.[6]

Retherford's results run counter to the prediction made in 1940 by Frederick Osborn, the first president of the Pioneer Fund. Osborn put forward what he called "the eugenic hypothesis" to the effect that dysgenic fertility in the United States and other economically developed nations would shortly be reversed.[7] Within a generation, Osborn believed, contraception would come to be used with virtually complete efficiency throughout the population. When this happened higher social classes, the better educated, and the more intelligent would have more children than the lower social classes, the poorly educated, and the least intelligent, because they would be better able to afford them. The more intelligent would have more children for the same reason that they have larger houses and more automobiles. Retherford's recent work has shown that Osborn was wrong when he predicted the imminent transition to eugenic fertility. On the contrary, dysgenic fertility has persisted into the 1990s.

NOTES

1. Retherford, R. D. & Sewell, W. H. 1988. Intelligence and family size reconsidered. *Social Biology*. 35: 1-40.

2. Retherford, R. D. & Sewell, W. H. 1989. How intelligence affects fertility. *Intelligence*. 13: 169-185.

3. Retherford, R. D., Ogawa, N. & Sakamoto, S. 1996. Values and fertility change in Japan. *Population Studies*. 50: 5-25.

4. Ogawa, N. & Retherford, R. D. 1993. The resumption of fertility decline in Japan: 1973-1992. *Population and Development Review*. 19: 703-741.

5. Retherford, R. D. 1994. Demographic transition and the evolution of intelligence. Unpublished paper.

6. Retherford, R. D. & Luther, N. Y. 1996. Differential fertility in relation to education in the United States. Unpublished paper.

7. Osborn, F. 1940. (1st ed.) *Preface to Eugenics*. New York, NY: Harper.

Roger Pearson, Ph.D.
Institute for the Study of Man

Chapter 35

Roger Pearson

Roger Pearson (born 1927) is an anthropologist whose steady stream of publications has provided a vital alternative to the equalitarian dogma that has held sway in the social sciences for most of the last 50 years. Pearson is publisher of *The Mankind Quarterly*, and founder and editor of *The Journal of Social, Political and Economic Studies*. As director of the Institute for the Study of Man, he has published and reprinted many important books on human heredity, variation, and evolution. Pearson is the author of *Race, Intelligence and Bias in Academe* which exposes the many media misrepresentations of behavioral science research and the politicizing nature of the equalitarian dogma.[1]

Roger Pearson was born in London, England in 1927, and educated in East Anglia. In 1945 he joined the British armed forces and was commissioned into the British Indian Army, serving in India and Malaya, and in the allied occupation of Japan. On leaving the Army in 1948, Pearson entered the University College of the South West of England in Exeter, where he graduated in economics, sociology, and anthropology. He took a master's degree at the University of London and began work on a Ph.D. at the London School of Economics, which he completed in 1969. In 1954 he published his first book, *Eastern Interlude*, a social history of the European community in Calcutta between 1689 and 1911.[2]

In 1954 Pearson went to India to work as an accountant in a British bank. He remained in India until 1965 and became director of several companies and chairman of the Pakistan Tea Association. It was during his years in India that Pearson became interested in genetics, heritability, eugenics, and race differences.

In 1965 Pearson resigned his business interests in India and took up an academic career. In 1967 he was appointed assistant professor of anthropology and sociology at the University of Southern Mississippi. In 1970 he moved to Queen's College in Charlotte, North Carolina, to become associate professor and head of anthropology and sociology. The next year he moved back to the University of Southern Mississippi as full professor and chairman of the department of anthropology. In 1974, Holt, Rinehart and Winston published his *Introduction to Anthropology*,[3] a textbook for university freshmen and sophomores, and in that same year he was appointed dean of academic affairs and director of research at the Montana College of Mineral Science and Technology. The next year he resigned his position to become director of the Institute for the Study of Man and of the Council for Social and Economic Studies, both located in Washington, D.C.

Pearson obtained his first grant from the Pioneer Fund in 1973 and since assuming the directorship of the Institute for the Study of Man that organization has been supported on an annual basis for academic work, publications, and the Institute library and running costs. From 1978 onwards Pearson has produced a steady stream of publications, including his *Anthropological Glossary*,[4] a dictionary of anthropological terms; *Race, Intelligence and Bias in Academe*,[5] an account of the politicization of the social and biological sciences by Marxists and politically-correct activists and their attacks on academic freedom; *Heredity and Humanity: Race, Eugenics and Modern Science*,[6] a history of eugenic ideas and an examination of modern genetics and the future of eugenics.

Pearson has also edited and co-authored a number of books, including *Korea in the World Today* (1977); *Sino-Soviet Intervention in Africa* (1977); *Essays in Medical Anthropology* (1981); *Ecology and Evolution* (1981); and *Shockley on Eugenics and Race* (1992), a collection of papers by William Shockley.

In addition to his academic work, Pearson has made an important contribution as a publisher. In 1975 he founded the *Journal of Social, Political and Economic Studies,* and assumed responsibility for the publication of the journal *Mankind Quarterly* in 1979. He has edited and published a number of monographs and supplements to these journals.

The two research organizations which Pearson heads have their own publishing houses, Scott-Townsend Publishers and the Cliveden Press, which have to date published more than 50 books. These include R. B. Cattell's *Intelligence and National Achievement* (1983), and *How Good is Your Country?* (1994); H. J. Eysenck's *The Decline and Fall of the Freudian Empire* (1990); Glenn Wilson's *The Great Sex Divide* (1992); Nathaniel Weyl's *The Geography of American Achievement* (1989); as well as reprinting such classics and other important books. Pearson has also used his publishing houses to reprint such classics of ethnology and eugenics as Sir Francis Galton's *Essays on Eugenics* (1909); Allen G. Roper's *Ancient Eugenics* (1913); David S. Jordan's *War and the Breed* (1915); Johannes Lange's *Crime and Destiny* (1929); and Wesley C. George's *Race Problems and Human Progress* (1967).

INTRODUCTION TO ANTHROPOLOGY

Between 1968 and 1973 Pearson worked on his 600-page college-level textbook, *Introduction to Anthropology,*[7] which he published in 1974. His book set out in detail the close relationship between biology and culture. His ecological/evolutionary approach emphasized the changing genotypes of species and subspecies as they face the challenge of survival in their environment and often modify that environment by their existence. Increasing intelligence and improved communication between individuals, Pearson argued, led to the accumulation of knowledge and learned patterns of behavior based on past experiences. This coalesced as culture. But culture rests fundamentally on the biological properties of the species or subspecies, its evolutionary *raison d'être* being to assist the members of the group to survive long enough to produce another generation, and to protect and educate that generation in order to pass on its genetic and cultural heritage, enriched, if possible, to future generations. Individuals are biologically important as the organisms which carry the genes, and selection plays upon differences in the distribution of genes among the individuals that comprise a population.

Selection also operates through competition between genetically divergent populations. Genetic improvements, reinforced by efficient cultural devices, can help a population to survive. Social behavior, too, is important. Ethics are rooted in the extent to which forms of behavior promote group survival. Loyalty, trustworthiness, and altruism contribute to the survival of the population. Defective genes can bring about the elimination of a population in the evolutionary struggle to survive, as can defective cultures. Dysgenic cultures can handicap a human population in the competitive struggle to procreate and gain access to the resources necessary for survival.

THE MANKIND QUARTERLY

In 1978 Pearson assumed responsibility for publishing the journal *Mankind Quarterly*. This journal was founded in 1960 by Robert Gayre, a Scottish anthropologist, who graduated at the University of Edinburgh, carried out post-graduate work at Oxford in the 1920s, and in the 1950s held an appointment as professor of anthropology at the University of Saugor in India. In the 1950s equalitarianism had established such a firm hold in the academic and publishing world that it had become virtually impossible to get hereditarian views published. As noted in Chapter 7, in the late 1950s Audrey Shuey was unable to find a publisher willing to publish her book *The Testing of Negro Intelligence*, which set out the evidence that blacks in the United States score lower than whites on intelligence tests and concluded that this difference had a genetic basis. Gayre decided that a journal was needed as a forum for the publication of papers presenting the hereditarian case on race differences and related issues and established *Mankind Quarterly* for this purpose.

The journal was supported by two honorary associate editors, Henry Garrett, chairman of the psychology department at Columbia University and a director of the Pioneer Fund from 1972-1973, and R. Ruggles Gates, F.R.S., professor of botany at King's College, London, and an eminent geneticist and eugenicist. The editorial board consisted of 24 international scholars, including Sir Charles Darwin, H. V. Vallois, Stanley Porteus, Frank McGurk, and Audrey Shuey. In the 1970s Hans

Eysenck and Richard Lynn joined the editorial board.

During the 18 years of Gayre's editorship, *Mankind Quarterly* served a useful purpose as an outlet for the publication of papers presenting the hereditarian view. The first number included papers by Sir Charles Darwin on the world population explosion,[8] R. Ruggles Gates on race differences in physiological characteristics and the effects of hybridization,[9] and Henry Garrett on race differences in intelligence in the United States.[10] Subsequent issues of the journal contained articles by a number of Pioneer grantees including Donald Swan, James Gregor, Robert Kuttner, Frank McGurk, and Richard Lynn.

When Pearson took over publication of the journal in 1978 he strengthened the editorial board by the addition of Hans W. Jurgens, head of anthropology at the University of Kiel, Germany; Peter Boev, an anthropologist at the University of Sofia, Bulgaria; Tadeusz Dzierzykray-Rogalski, chairman of the Polish Anthropological Society; Brunetto Chiarelli, geneticist and anthropologist, and head of anthropology at the University of Florence, Italy; Raymond B. Cattell, an expert on intelligence and personality testing; Volkmar Weiss, head of the German genealogical institute in Leipzig; and Umberto Melotti, head of anthropology at the University of Pavia, Italy. Later the editorial board was further strengthened by the addition of John L. Horn of the University of Denver; Seymour Itzkoff of Smith College; Balslev Jorgensen of the University of Copenhagen; David de Laubenfels of Syracuse University; Clyde Noble of the University of Georgia; Daniel Vining of the University of Pennsylvania; Edgar Polomé of the University of Texas; and Joseph Campbell, the renowned expert on world mythology. Under Pearson's direction, *Mankind Quarterly* has continued to publish important papers on anthropology, psychology, and the sociology of evolution, genetics, eugenics, intelligence, personality, and race differences.

RACE, INTELLIGENCE AND BIAS IN ACADEME

In 1991 Pearson produced *Race, Intelligence and Bias in Academe*, an account of the assault by political activists and equalitarians on scientific inquiry into the issues of the heritability of intelligence and of social class and race differences.[11] Pearson argued that the mainline scientific research had shown that intelligence has a high heritability and that there is a genetic basis to the social class and race differences in intelligence. Nevertheless, these conclusions have been violently attacked by a small group of vocal academics, prominent among whom have been Leon Kamin, Stephen Jay Gould, and Barry Mehler.

The book gives an account of how the politically motivated left has campaigned to suppress the truth on these issues. Leftist and Marxist academics have misrepresented the evidence, and radical activists have intimidated and physically attacked academics who have taken the hereditarian position. Pearson gives a detailed account of such attacks on Jensen, Eysenck,

Shockley, Rushton, and a number of other leading academics who have written and lectured in support of the hereditarian case. These attacks have succeeded in intimidating many academics to the extent that few dared voice their real views on these questions, academic textbooks misrepresent the truth, and university administrators often fail to support those few academics who have the integrity and courage to speak out on these issues.

HEREDITY AND HUMANITY

Pearson's most recent book, *Heredity and Humanity: Race, Eugenics and Modern Science*,[12] gives an account of how the importance of heredity as a determinant of human intelligence and personality was well understood in the classical ancient Europe and in western society up to the end of the 19th century and into the first half of the 20th, but how in the second half of the 20th century this view has come increasingly under attack by politicized writers. Pearson restates his conclusion that this attack has been primarily motivated by Marxist ideologues and their fellow travelers and that "the political left wing has now achieved ascendancy in the universities of the Western world."[13]

The body of the book is a history of eugenics beginning with the ideas of Sir Francis Galton and the widespread acceptance of eugenics in England in the early decades of the 20th century by people like Karl Pearson, H. G. Wells, G. B. Shaw, Sidney and Beatrice Webb, John Maynard Keynes, Julian Huxley, and Havelock Ellis. The ideas spread to the

United States and Europe. In the United States the Eugenics Record Office was established in 1910 by Charles Davenport and administered by Harry Laughlin, the first president of the Pioneer Fund. Many leading biologists and social scientists in America subscribed to eugenics, including the psychologists Robert Yerkes, William McDougall, Emory Bogardus, and Lewis Terman, and the sociologists Franklin Giddings, Edward Ross, and Frank MacIver. Eugenicists were often committed conservationists, seeking to preserve the environmental as well as the genetic heritage of their generation.

Pearson sees the assault on hereditarian and eugenic thinking in the United States as being led in the first half of the 20th century by Franz Boas, professor of anthropology at Columbia University, whose 1940 book *Race, Language and Culture* disputed that genetic factors are involved in individual, social class, and racial differences. Boas built up an influential school of anthropology at Columbia and was assiduous in advancing the careers of those of his pupils who shared his beliefs so that they could secure dominant positions in American anthropology.

As an anthropologist, Pearson has also concentrated on the sociological aspects of human evolution in such books as his *Ecology and Evolution* and *Introduction to Anthropology*,[14] exploring the link between biology and culture with special reference to the eugenic and dysgenic aspects of the socio-cultural factors influencing man's existence.

NOTES

1. Pearson, R. 1991. (1st ed.) *Race, Intelligence and Bias in Academe.* Washington, D.C.: Scott-Townsend; 1997 (2nd ed.).

2. Pearson, R. 1954. *Eastern Interlude.* Calcutta: Thacker Spink.

3. Pearson, R. 1974. *Introduction to Anthropology.* New York, NY: Holt, Rinehart and Winston.

4. Pearson, R. 1985. *Anthropological Glossary.* Washington D.C.: Scott-Townsend.

5. Pearson, R. 1997. *Op cit.*

6. Pearson, R. 1996. *Heredity and Humanity: Race, Eugenics and Modern Science.* Washington, D.C.: Scott-Townsend.

7. Pearson, R. 1974. *Op. cit.*

8. Darwin, C. G. 1960. World population. *Mankind Quarterly.* 1: 5-10.

9. Gates, R. R. 1960. The emergence of racial genetics. *Mankind Quarterly.* 1: 11-14.

10. Garrett, H. E. 1960. Klineberg's chapter on race and psychology. *Mankind Quarterly.* 1: 15-22.

11. Pearson, R. 1997. *Op cit.*

12. Pearson, R. 1996. *Heredity and Humanity: Race, Eugenics and Modern Science.* Washington, D.C.: Scott-Townsend.

13. Pearson, R. 1981. *Ecology and Evolution.* Washington, D.C.: Cliveden Press. 73.

14. Pearson, R. 1974. *Op. cit.*

Hiram P. Caton, Ph.D.
Griffith University

Chapter 36

Hiram P. Caton

Hiram Caton (born 1936) is a political scientist and ethnologist interested in the biological basis of human political and social behavior. He has made notable contributions in the controversy over the anthropological work of Margaret Mead, in the compilation of bibliographies of human behavior, in the analysis of collective behavior, and in the future of genetic engineering.

Hiram Caton, Professor of Politics and History at Griffith University in Brisbane, Australia, has worked forcefully and comprehensively to bring biology back into the social sciences. A political scientist and ethnologist, Caton edited *The Samoan Reader,*[1] a collection of essays critical of Margaret Mead's classic research in Samoa, which had long been the standard text allegedly proving the massive malleability of human nature. *The Bibliography of Human Behavior,*[2] which Caton edited with Frank Salter and J. M. G. van der Dennen, is the most extensive and up-to-date reference guide on comparative human social behavior. As head of the School of Applied Ethics, Caton has also studied the potential impact of state-of-the art technologies such as amniocentesis and ultrasound on reproductive policy. His most recent research examines crowd behavior from an evolutionary, rather than a psychotherapeutic, point of view.

Hiram P. Caton was born in Concord, North Carolina, in 1936 and educated at the University of Chicago where he took his B.A. in 1960 in Arabic Language and Civilization at the Oriental Institute. He obtained an M.A. at Chicago in 1962 and a Ph.D. in philosophy at Yale in 1966. From 1966-1971 he taught in the philosophy department at Pennsylvania State University. He emigrated to Australia to work in the Research School of Social Sciences at the Australian National University. From 1976 he has been head of the School of Applied Ethics and Professor of Politics and History at Griffith University in Brisbane, Australia. In 1988 he was elected a Fellow of the Australian Institute of Biology. Caton's work has been supported by the Pioneer Fund from 1990 onwards.

THE SAMOAN READER

In 1990 Caton edited *The Samoan Reader*, a collection of essays on the Margaret Mead controversy.[3] Margaret Mead was one of the most influential American anthropologists of the 1920s and 1930s. She graduated at Columbia University as a student of Franz Boas, the leader of the cultural relativism school of anthropology, which argued against the existence of biologically programmed psychological universals such as ethnocentrism, male competitiveness, and so forth, and maintained that human behavior is infinitely variable and malleable in different cultures.

Boas gave Mead the assignment of carrying out field work on the south Pacific island of Samoa, in which she was to contrast the uninhibited life style of a primitive people to the repressed culture of the United States. In particular, she was to seek evidence to challenge the contention of G. Stanley Hall that adolescent rebellion expressed developmental endocrine changes underlying sexual maturation. Hall proposed his explanation to account for violent delinquent juvenile gangs. Mead obliged Boas by carrying out the work and writing a book, *Coming of Age in Samoa*,[4] in which she depicted adolescence in Samoa as free from the emotional stress, sexual jealousy, and conflicts with parents, siblings, and society present in economically developed nations. She ascribed this happy outcome to the absence of repressive upbringing and of sexual restraint. She promoted Samoa as a model for permissive child rearing and urged that the allegedly uninhibited Samoan life style should be adopted as a model by the United States.

In 1983 Mead's work was critically examined by Daniel Freeman in his book *Margaret Mead and Samoa*.[5] Freeman compared Mead's descriptions of life in Samoa with the ethnographic evidence. He found that, contrary to Mead's reports, Samoan nurturing is repressive. Corporal punishment is frequent from infancy onwards, extra-marital sexual relations are prohibited in Samoan mores, and Samoan men are aggressive towards other males and towards women. In short, Mead's description of life on the island paradise was wrong from start to finish. Freeman also documented in his book the steps by which the Boasian school disengaged anthropology from its links with biology and made biological approaches to behavior taboo among American anthropologists. Freeman's book generated a controversy in which a number of cultural anthropologists defended Mead's work, while others supported Freeman. Caton's edited volume, and his own contributions to it, are a source book for this controversy.

EUGENIC POTENTIAL OF GENETIC ENGINEERING

Caton has written about the eugenic possibilities of genetic engineering.[6] He describes the advancement during the last quarter of the 20th century in the diagnosis of genetic and congenital fetal abnormalities by the use of amniocentesis, blood tests, and ultrasound, and the legalization of abortion in most economically developed nations in cases where abnormalities are identified. This is negative eugenics, in so far as it reduces the prevalence of harmful genes in the gene pool.

Caton foresees the possible future development of genetic engineering to promote positive eugenics. He suggests that a possible scenario is that *in vitro* fertilization will become increasingly used to screen embryos for genetic disorders. Several embryos will be grown in test tubes and genetically assessed, initially for genetic diseases and then for personality traits, minor physical disabilities, and appearance. He does not mention intelligence, but no doubt it will be possible in due course to assess the fetus for this as well. Prospective parents could then choose which of a number of fetuses to implant in the mother, while the less desirable would be discarded.

Caton sees a development of this kind as an extension of parental choice and as ethically acceptable. He argues that parents can already choose not to have a genetically impaired child. Since this is already permissible, he sees no reason why parents should not be allowed to chose to have babies with other qualities.

Caton also suggests that genetic engineering by *in vitro* fertilization and selective implantation is likely to become increasingly used because of cost considerations. The rearing of a seriously genetically impaired child entails considerable costs, which have to be borne either by the parents or by society. With medical advances in the diagnosis of genetically impaired fetuses, and the increasing use of abortion following a positive diagnosis, bearing a genetically impaired child will come to be regarded as negligent, and the cost will come to be imposed on the parents. This will provide an incentive for these parents to have defective fetuses aborted.

BIBLIOGRAPHY OF HUMAN BEHAVIOR

In the mid-1980s Caton began work on a bibliography of human behavior, with the assistance of Frank K. Salter and J. M. G. van der Dennen. This project was supported by the Pioneer Fund and the Max Planck Institute for Human Ethology in Germany. *The Bibliography of Human Behavior*[7] lists over 6,700 titles arranged under 20 headings, among them human evolution, prehistory, population and human biology, cultural evolution, sociobiology, behavior genetics, ethology, endocrinology, neurology, emotion, cognition, social psychology, and psychiatry. It is the most comprehensive bibliography of its kind, selected from over 160,000 titles.

PEOPLE POWER: A STUDY OF COLLECTIVE ACTION

Caton's current project, supported by the Pioneer Fund, is an ethologically-oriented study of crowd psychology and behavior.[8] His point of departure is the work of the early French sociologist Gustave LeBon who proposed that crowds have a psychological unity maintained by four mechanisms: imitation, suggestion, contagion, and de-individuation. Caton believes that ethnological research on the development of infant behavior shows that imitation is the fundamental process by which infants learn. He argues that LeBon's four mechanisms are not peculiar to crowds but are present in all human interaction, and particularly in regimented mass action, for instance in military units or corporate teams. LeBon and the sociological tradition of collective behavior tacitly identified the crowd with the rioting mob and the mob as irrational, emotional, and violent. Caton believes that this overlooks the fact that rioting is often a deliberately selected conflict strategy, developed historically by trade unions and refined by Bolsheviks. It also ignores the fact that most politics of the streets is peaceful, and is sponsored by organizations who use it to advance political objectives. For example, one of the most influential protest strategies of this century, civil disobedience, involves a nonviolent challenge to authority.

To identify what is special about crowd politics, Caton arranges crowds in a continuum from mobs through protest crowds, sporting and entertainment crowds, ceremonial crowds, and festivals. The common thread is a ritual celebration of group solidarity. He maintains that the rioting mob is not a nondescript frenzied mass. It executes an identifiable group's retribution against an outgroup. Lynching, communal violence in Asia, and European revolutionary violence are all expressions of this model. Ceremonial and festive crowds, particularly those sponsored by governments or religious groups, reaffirm myths explaining and justifying the group identity. Across this whole continuum Caton finds auto-manipulative techniques by which crowds and crowd leaders trigger pro-social behavior and feelings commonly called "community" and "solidarity."

"De-individuation" is the effect of the brain's reward system which, triggered by auto-manipulation, transports crowds to elevated, ecstatic mood states whose expression is readily observed in sports crowds. These mood states are typically characterized by strong feelings of a tribal "them-us" dichotomy. At one and the same time they are intimidating to bystanders or opponents and have euphoric effects on group members. The euphoria stimulates heroic struggle and acts of self-sacrifice. It organizes the neural networks that support killing of outgroup members, irrespective of age or sex.

Caton believes that these euphoric states have a genetic basis derived from human evolutionary origins. In support of this view he notes that violence is overwhelmingly committed by young males; that military units and police have been all male until very recently; and that rioters arrested for violent offenses, such as assault and arson, are typically 85-95 percent young males.

Caton's study conceptualizes police crowd control as a form of collective behavior. He notes that crowd events are comprised of the crowd, police, and bystanders and that crowd events are the outcome of the interactions between these three elements. He believes that the failure to observe that police are integral to crowd events is a blunder equaling the failure to observe that crowd behavior varies with the age and sex of its composition. Police are young males directed by elders because, in evolutionary history, they are the order-imposing element of human association.

Although the orientation of his study is ethnological, Caton incorporates neurology in his theory of the behavioral synchrony of groups. He believes that the neurology of sensory information processing makes it possible to explain the contribution of evolved basic structures and the happenstance quality of actual group identity to the behavior of crowds and groups. At the time of this writing Caton is working on this set of problems.

NOTES

1. Caton, H. (ed.) 1990. *The Samoan Reader: Anthropologists take Stock.* Cambridge, MA: Harvard Univ. Press.

2. Caton, H., Salter, F. K. & van der Dennen, J. M. G. 1993. *Bibliography of Human Behavior.* Westport, CT: Greenwood Press.

3. Caton, H. (ed.) 1990. *Op. cit.*

4. Mead, M. 1928. *Coming of Age in Samoa.* New York: Morrow.

5. Freeman, D. 1984. *Margaret Mead and Samoa: The Making and Unmaking of an Anthropological Myth.* New York, NY: Penguin.

6. Caton, H. 1992. Some social consequences of gene improvement. *Policy.* 8:29-32.

7. Caton, H., Salter, F. K. & van der Dennen, J. M. G. 1993. *Op. cit.*

8. Caton, H. 1995. Selective witness to a century of biology and culture. *Journal of Social and Evolutionary Systems.* 18: 103-107.

Richard Lynn, Ph.D.
University of Ulster

Chapter 37

Richard Lynn

Richard Lynn (born 1930) is a British psychologist who is best known for his work showing that the Asian peoples of China and Japan have slightly higher average IQs than Caucasians in the United States and Europe. He has also worked on the intelligence of populations as a determinant of their economic and intellectual achievement, the dysgenic effects of immigration and emigration, the evolution of race differences in intelligence, race differences in reaction times, the role of nutrition in the secular increase in intelligence, sex differences in brain size and intelligence, and dysgenic trends in modern populations.

Richard Lynn was born in 1930 and brought up in Bristol, England. His father was a plant breeder, from whom Lynn learned that genes are an important determinant of the quality of plants, and Lynn never found any difficulty in extending this principle to humans. Lynn was educated at the

Bristol Grammar School. He spent 1948-49 as a second lieutenant in the British Army. In the fall of 1949 he went up to the University of Cambridge, where he graduated with a B.A. in psychology in 1953 and obtained his Ph.D. in 1956. He spent the years 1956-1967 as a lecturer in psychology at the University of Exeter. In 1967 he moved to the Republic of Ireland to become professor of psychology at the Dublin Economic and Social Research Institute. In 1972 he was appointed professor of psychology at the University of Ulster in Northern Ireland, where he has spent the remainder of his career.

In the early years of his career Lynn worked in physiological psychology and personality theory. He was drawn to physiological psychology through a belief that physiological processes are the basis of psychological characteristics and need to be understood to advance the understanding of psychology. In the mid-1960s he wrote his first book, *Arousal, Attention and the Orientation Reaction*,[1] on the neurophysiological processes involved when people pay attention to novel stimuli and gradually cease to notice stimuli as they become familiar.

In personality, Lynn's most significant work was his theory that it is possible to measure national differences in the trait of anxiety from a combination of demographic indices such as the rates of suicide, alcoholism, accidents, and so forth.[2] Lynn argued that these could be envisaged as functions of the level of anxiety in national populations. His analysis showed that Japan and the nations of southern and central Europe had high levels of anxiety, while among the northern nations of Britain and Scandinavia the level of anxiety was low. This led him to wonder about the causes of these national differences and whether there might be genetically based racial factors responsible for differing national anxiety levels. It set him thinking about the possibility that there might be racial differences in other psychological traits. In the mid-1970s Lynn began to consider this question in relation to intelligence. His work on this and related problems has been supported by the Pioneer Fund since 1983.

THE SOCIAL ECOLOGY OF INTELLIGENCE

The first problem Lynn took up in the field of intelligence concerned the question of whether the intelligence level of a population contributed to its intellectual achievements. Sir Francis Galton had assumed that this was the case and had argued that the people of classical Greece must have had an exceptionally high level of intelligence to produce their great works of mathematics, astronomy, philosophy, and literature. However, it had never been demonstrated that there was an association between the intelligence level of a population and its intellectual achievements. Lynn first tackled this problem by calculating average IQs for 13 regions of the British Isles.[3] He showed that the average IQs ranged from 102 in London to 97 in Scotland and 96 in Ireland. He showed that the intellectual achievement of the regions indexed by the numbers of Fellows of the Royal Society ran parallel to the mean IQs, being 8.7 per million born in London and 2.5 born in Scotland. Thus, a difference of 5 IQ points between the populations of London and Scotland produces a more than threefold difference in the proportions of Fellows of the Royal Society produced by the two regions.

Lynn also showed that the regional differences in per capita income and unemployment were strongly related to the differences in average intelligence, per capita income being highest in London and unemployment lowest. Finally, he considered the question of the factors responsible for the regional differences in intelligence levels. He concluded that they had come about as a result of the migration of more intelligent individuals from the provinces to London over a period of centuries, and that this had impoverished the gene pool of the provinces and enhanced that of London.

Lynn next made a related study of the regional differences in intelligence in France.[4] He found a similar pattern, the average IQ being highest in Paris and lowest in the more remote provinces; he found the same regional differences in intellectual achievement, indexed by membership of the Academie Français and per capita income, which were both greatest in Paris. He concluded that long term selective migration of the more intelligent individuals from the provinces to the capital city had operated in France in the same manner as in the British Isles. More generally, the results of the two studies illustrated the eugenic impact of migration, in so far as London and Paris had benefitted from eugenic inward migration, while the provinces had suffered.

As Lynn was securing evidence that regional differences in intelligence in the British Isles and in France contributed to differences not only in intellectual achievement but also in incomes, he began to wonder whether intelligence levels might be a determinant of differences in economic performance among nations. It seemed probable that the populations of different nations would have different average levels of intelligence and that this would be a determinant of economic performance, just as it appeared to be for the different regions of Britain and France.

Lynn began to think in particular about the rapid economic growth of Japan, which was becoming recognized in the 1970s as the economic miracle of the post World War II decades, and he started to wonder whether the Japanese might be unusually intelligent and whether this might be a factor in their strong economic performance. At this time nothing was known of the intelligence of the Japanese, but in 1977 Lynn hit upon a method for measuring it. This consisted of making calculations from the Japanese standardizations of the American Wechsler tests. The results indicated that the Japanese had an average IQ of 106.6, as compared with a mean of 100 for American Caucasians.[5] It appeared that Lynn's hunch that the Japanese are an exceptionally intelligent people was correct.

Lynn began to look next for evidence on the intelligence levels of other East Asian populations.

In 1977 he discovered an unpublished Ph.D. thesis containing data on the performance of the Chinese in Singapore on the Progressive Matrices Test, from which he calculated that their mean IQ was 110.[6] The next year he published a general review of national and racial differences in intelligence in which he found more evidence that Oriental peoples generally have average IQs somewhat higher than those of American and European Caucasians. He also showed that the low IQ of blacks is not confined to the United States, but is also present among blacks living in Britain, in the Caribbean, and in Africa.[7] He argued that the universality of the differences in IQ levels between the different races corroborates other evidence suggesting that they have a genetic basis.

Over the next two decades Lynn has carried out a number of further studies of the intelligence of the East Asian peoples in Japan and also in Hong Kong, Korea, Taiwan, and the United States. In 1987 he published a general review of the evidence which indicated that these East Asian peoples invariably have average IQs a little higher than those of North American and European Caucasians. These IQs typically lie in the range of 102-108.[8] In 1996 he published the results of a study of Americans of East Asian origin, in which he calculated their IQs at 104.4.[9] Lynn has also found that in addition to their high average IQs, East Asians have a distinctive pattern of intelligence consisting of exceptionally strong spatial abilities and rather weaker verbal abilities. He has found that this pattern is present among East Asians and their descendants wherever they are situated and shows up in the United States in their stronger performance on the math section of the Scholastic Aptitude Test than on the verbal.

In addition to his work on the high average IQ of the Japanese, Lynn has worked with a Japanese colleague, Ken Hattori, on the heritability of intelligence in Japan. In a study of 543 identical 12 year old twins and 134 non-identicals they found correlations for intelligence of 0.78 and 0.49, respectively, indicating a heritability of 58 percent.[10]

In the later 1980s, Lynn extended his work on race differences in intelligence to include research on the question of whether there are race differences in reaction times. The use of reaction times as a measure of the neurological efficiency of the brain had been developed by Arthur Jensen, Hans Eysenck, and Tony Vernon. Lynn carried out a series of studies designed to determine whether there are racial differences in reaction times parallel to those of the racial differences in intelligence, as measured by intelligence tests. He organized studies of reaction times and intelligence in children in Britain, Ireland, Japan, Hong Kong, and blacks in South Africa. The results confirmed previous studies showing that there is a positive correlation between reaction times and intelligence, and showed further that reaction times were fastest among children in Japan and Hong Kong; intermediate among children in Britain and Ireland; and slowest among South African children.[11]

The reaction time studies showed that racial differences in intelligence measured by tests are also present in these measures of the neurophysiological efficiency of the brain. This overcame the criticism made by equalitarians that intelligence tests are biased against blacks by showing that the same racial ordering appears even on simple tests of neurophysiological functioning. It provided a further strand of evidence for a genetic component to the race difference.

THE EVOLUTION OF RACE DIFFERENCES IN INTELLIGENCE

In 1991 Lynn produced a review of the evidence on race differences. It showed that Caucasians in North America, Europe, New Zealand, and Australia obtain mean IQs of about 100; East Asians typically obtain slightly higher mean IQs; African Negroids average around 70; Negroid-Caucasian hybrids in the United States and Britain average around 85; and Amerindians and Southeast Asians average around 90.

In the 1980s, as Lynn collected data showing that East Asians have higher average IQs than Caucasians, he began to think about why the East Asian peoples had evolved their high IQs and their distinctive pattern of high spatial abilities. His solution to this problem was that these abilities had evolved in response to the selection pressure of the intense cold of northeast Asia during the last ice age, which lasted approximately from 28,000 years ago to 12,000 years ago.

Lynn proposed that during this cold period the Oriental peoples would have had to build shelters, make clothes, and keep fires burning in order to stay warm and hunt large animals to secure their food supply during the winter and spring when no plant foods were available. The hostile climate would have acted as a selection pressure for enhanced intelligence, especially of the spatial abilities required for hunting large prey and for constructing the flint tools necessary for killing and butchering them. Those with low intelligence would not have been able to survive in the harsh northern environment and would have been eliminated from the population. Lynn extended this climatic theory of the origin of the high intelligence of the Orientals into a general theory of the evolution of race differences in intelligence.[12]

THE SECULAR INCREASE OF INTELLIGENCE

In 1982 Lynn showed in one of his papers on the intelligence of the Japanese that the average level of intelligence in Japan had risen quite significantly since the 1930s.[13] To pursue this striking discovery, he investigated whether intelligence has also been increasing in Britain and reported in 1987 that the intelligence of British 10 year olds had increased by 12.5 IQ points over the half century from 1936-1986.[14] These studies stimulated James Flynn to collect data on the extent to which intelligence has been rising in other industrial nations and he concluded that this has been the case throughout the economically developed world.[15]

These discoveries raised the problem of what factors have been responsible for these increases in intelligence that have apparently occurred during the course of the 20th century. Most psychologists thought that they were due to improvements in education, but Lynn assembled an array of different kinds of evidence which pointed to improvements in nutrition as by far the most important factor .[16] In a further study of the effects of nutrition on intelligence, Lynn and his colleagues have shown that the administration of iron supplements to anemic children over a three month period increases their intelligence by approximately 5 IQ points.[17]

BRAIN SIZE, INTELLIGENCE AND RACE DIFFERENCES

One of the arguments Lynn advanced in support of his theory that improvements in nutrition have been the principal factor responsible for the secular increases in intelligence was that one determinant of intelligence is brain size, and average brain size in the industrialized nations has increased during the 20th century as a result of improvements in nutrition. The thesis that brain size is a significant determinant of intelligence has often been disputed but Lynn was able to assemble a number of studies which had found this and has confirmed that there is a positive association between brain size and intelligence in two studies of his own, one carried out in Ireland and the other in India.[18]

Lynn has extended the relationship between brain size and intelligence to the issue of race differences in intelligence and has shown in two studies that in the United States blacks have smaller average brain size than whites, even when body size is taken into account.[19] He argues that the smaller average brain size of blacks is one factor responsible for their lower average intelligence.

SEX DIFFERENCES IN BRAIN SIZE AND INTELLIGENCE

The theory that differences in brain size are a determinant of intelligence encountered the problem of the difference in the brain size of males and females. Males have larger brains than females, yet it was universally asserted in psychological textbooks that males and females have the same average intelligence. There was evidently an anomaly here which required resolution.

In 1992-1993 Lynn worked on this problem and proposed the solution that, among adults, males do have higher average intelligence than females by approximately 4 IQ points.[20] He argued that hitherto psychologists had failed to spot this sex difference because it is not present in children, as a result of the earlier maturation of girls, and only appears at about the age of 16 years; and because males have a considerable advantage over females in spatial abilities, which are underrepresented in intelligence tests. This explanation resolved the anomaly and strengthened further the theory that brain size is a significant correlate of intelligence and explains some of the sex differences in intelligence.

DYSGENICS

In 1994 Lynn undertook a study of the eugenic concern that modern populations are deteriorating genetically in respect of the increased incidence of genetic diseases, and in regard to their intelligence and moral character. In his book, *Dysgenics*,[21] he assembled the evidence in support of this view. He shows that medical advances in the treatment of a large number of genetic diseases have enabled many of those afflicted to survive and have children, thereby transmitting the genes for these conditions to future generations. He shows that throughout virtually the entire world intelligent individuals are having fewer children than the unintelligent, which he attributes largely to their more efficient use of contraception; and that this must entail a deterioration in the genetic quality of the world population for intelligence, although this has been masked in the industrialized countries by improvements in nutrition.

Finally, with regard to personality, he has documented a wide range of evidence showing that this has a genetic basis and that more conscientious parents are having fewer children than those whose moral character is weak; and that, in particular, criminals have substantially greater numbers of children than the law-abiding. Lynn's book argues that eugenicists were right in their concerns about the genetic deterioration of modern populations and that conscious eugenic measures need to be taken to arrest and reverse these trends.

In 1997, Lynn returned to the issue of race differences in intelligence and produced an updated review of the world literature. This confirmed his previous conclusion that the average IQs were: East Asian (105), European Caucasians (100), African Americans (85), and Sub-Saharan Africans (70).[22]

Despite its controversial nature, Lynn's work has received professional recognition. He has been elected a Fellow of the British Psychological Society and of the Galton Institute, and he has received the American Mensa Award for Excellence in 1989 for his work on the high intelligence of the Orientals and again in 1993 for his work on the role of nutrition in the secular increase of intelligence. Like a number of other social and biological scientists sponsored by the Pioneer Fund, his lectures have been picketed and disrupted and leftist student activists have demanded his dismissal.

NOTES

1. Lynn, R. 1967. *Arousal, Attention and the Orientation Reaction.* Oxford, U.K.: Pergamon Press.

2. Lynn, R. 1971. *Personality and National Character.* Oxford, U.K.: Pergamon Press.

3. Lynn, R. 1979. The social ecology of intelligence in the British Isles. *British Journal of Social and Clinical Psychology.* 18: 1-12.

4. Lynn, R. 1980. The social ecology of intelligence in France. *British Journal of Social and Clinical Psychology.* 19: 325-330.

5. Lynn, R. 1977. The intelligence of the Japanese. *Bulletin of the British Psychological Society.* 30: 69-72.

6. Lynn, R. 1977. The intelligence of the Chinese and Malays in Singapore. *Mankind Quarterly.* 18: 125-128.

7. Lynn, R. 1978. Ethnic and racial differences in intelligence: international comparisons. In R. T. Osborne, C. E. Noble and Weyl, N. (eds.) *Human Variation.* New York, NY: Academic Press.

8. Lynn, R. 1987. The intelligence of the Mongoloids: a psychometric, evolutionary and neurological theory. *Personality and Individual Differences.* 8: 813-844.

9. Lynn, R. 1996. Racial and ethnic differences in intelligence in the United States on the Differential Ability Scale. *Personality and Individual Differences.* 20: 271-273.

10. Lynn, R. & Hattori, K. 1990. The heritability of intelligence in Japan. *Behavior Genetics.* 20: 545-546.

11. Lynn, R. 1991. Race differences in intelligence: a global perspective. *Mankind Quarterly.* 31: 254-296.

12. Lynn, R. 1991. The evolution of racial differences in intelligence. *Mankind Quarterly.* 32: 99-173.

13. Lynn, R. 1982. IQ in Japan and the United States shows a growing disparity. *Nature.* 297: 222-223.

14. Lynn, R., Hampson, S. & Mullineux, J. 1987. A long term increase in the fluid ability of English children. *Nature.* 328: 797.

15. Flynn, J. 1987. Massive IQ gains in 14 nations: what IQ tests really measure. *Psychological Bulletin.* 101: 171-191.

16. Lynn, R. 1990. The role of nutrition in the secular increases in intelligence. *Personality and Individual Differences.* 11: 273-285.

17. Lynn, R. & Harland, P. E. 1998. A positive effect of iron supplementation on the IQs of iron deficient children. *Personality and Individual Differences* 24: 883-885.

18. Lynn, R. 1990. New evidence on brain size and intelligence: a comment on Rushton and Cain and Vanderwolf. *Personality and Individual Differences.* 11: 795-797; Lynn, R. and Jindal, S. 1994. Positive correlations between brain size and intelligence: further evidence from India. *Mankind Quarterly.* 34: 109-115.

19. Lynn, R. 1990. *Op. cit.*; Lynn, R. 1993. Further evidence for the existence of race and sex differences in cranial capacity. *Social Behavior and Personality.* 21: 89-92.

20. Lynn, R. 1994. Sex differences in intelligence and brain size: a paradox resolved. *Personality and Individual Differences.* 17: 257-271.

21. Lynn, R. 1996. *Dysgenics.* Westport, CT: Praeger.

22. Lynn, R. 1997. The geography of intelligence. In H. Nyborg (ed.) *The Scientific Study of Human Nature.* Oxford, U.K.: Pergamon.

Linda S. Gottfredson, Ph.D.
University of Delaware

Chapter 38

Linda S. Gottfredson

Linda Gottfredson (born 1947) is a sociologist who has broken ranks by recognizing the importance of intelligence in job performance and the significance of racial differences in intelligence. Her early work concerned traditional questions in the sociology and psychology of career development and vocational counseling. Over time her work evolved to focus on public policy issues regarding education and employment, including fairness in mental testing, the politicization of science with regard to race, and the sociopolitical ramifications of differences in intelligence, both within and between racial and ethnic groups.

Linda S. Gottfredson was born in 1947 and brought up near the town of Davis, California, where both her father and grandfather worked as professors of veterinary medicine at the University of California's agricultural campus. She entered the University of California at Berkeley, where she completed a bachelor's degree in psychology in 1969. It was during her years at Berkeley, a time of national turmoil and local race riots, that she began turning her attention to issues of racial inequality. She worked as a volunteer teacher's aide in an inner-city school in Oakland and for the Alameda County Human Relations Commission, helping to advertise its services and examining racial bias in the profession of journalism. From 1969 to 1972 she worked as a Peace Corps volunteer at the Ministry of Health in Malaysia.

Soon after returning to the United States, Gottfredson enrolled as a graduate student in sociology at Johns Hopkins University, where she obtained her Ph.D. in 1977. There she specialized in the field of social stratification, which focused at that time on mathematically modeling social mobility. She sought in her dissertation to improve the standard sociological models by integrating vocational psychology's focus on field and functional demands of work with sociology's focus on work's different levels and rewards.

INTELLIGENCE AND JOB PERFORMANCE

For the next decade Gottfredson worked with a team of sociologists at Johns Hopkins, funded by the federal government, to investigate the ways in which schooling reduces or increases educational and occupational inequalities. Much of her initial work was concerned with the problem of how to improve vocational counseling and career opportunities for women, minorities, people with handicaps, and other special groups. In 1981 she published a monograph in which she drew psychologists' attention to the social and cognitive processes by which children and young adolescents progressively narrow their own occupational aspirations according to self-conceptions of gender, social class, and academic ability, long before they encounter actual barriers in the labor market.[1]

In the late 1970s increasingly strong claims were being made that ability tests and counselors were culturally biased and this had made vocational psychologists reluctant to raise issues of ability, hitherto a traditional concern in the field, with their counselees. Gottfredson believed that ability was important and that vocational counselors needed to devote more attention, not less, to job-aptitude demands if they wished to help counselees improve their chances of success. She believed that to be able to do this counselors would need more systematic knowledge about the skills and abilities required in different careers and better ways of introducing the information to counselees. She located a wealth of data on job-aptitude requirements, mostly in personnel selection psychology, which she analyzed to reveal general patterns of the aptitude demands in different fields of work. These she distilled into an easily understandable occupational-aptitude patterns map. Counselors could use the map to encourage counselees not only to be more realistic about their relative competitive advantage, but also to reconsider viable options they might have rejected earlier as seemingly inconsistent with their gender or social class.

The profile of job-aptitude requirements that Gottfredson discovered was not at all what she had been led to expect by her training in sociology or her reading in counseling psychology. Contrary to claims common in these fields, she concluded from the research literature that intelligence is the most important measurable factor in the world of work. Furthermore, while useful in all jobs, intelligence becomes critical in the more complex and highly prestigious occupations. It predicted job performance better than any other single trait or circumstance, including education and specific aptitudes. This knowledge was of great value because technological changes have made high-level skills increasingly important, even in jobs where formerly the unskilled could cope. The growing complexity of western civilization has been raising the level of intelligence required in the labor force. Gottfredson concluded that the occupational-prestige hierarchy is essentially a scale that orders occupations according to their intellectual difficulty.

Gottfredson was also not prepared by her training to discover in the psychological literature that the average IQs of racial-ethnic groups differ, that (whatever their cause) the differences were large, and that they have been constant during the course of the 20th century. Yet much public policy with regard to race in the United States was based on what she would later term "the egalitarian fiction," namely the assertion that there exist no meaningful differences in intelligence between racial-ethnic groups. She began to explore the ramifications of individual and group differences in intelligence in relation to the prevailing theories of social inequality. She published a major theoretical synthesis of the empirical literature in job analysis, personnel selection, and psychometrics, in which she argued that the common occupational-prestige hierarchy that sociologists have demonstrated in numerous countries is a function of differences in intelligence.[2] In the 1980s Gottfredson continued to work on the practical consequences of racial and ethnic differences in intelligence for educational and vocational achievements. In 1986 and 1988 she edited two issues of the *Journal of Vocational Behavior*, in which she assembled a variety of experts to discuss this issue and to which she herself contributed articles.[3]

RACE NORMING

In 1990 Gottfredson took up the issue of the "race norming" of intelligence test scores for hiring policies. The background to this question was that intelligence tests had long been used in personnel selection because it was established that IQ is a determinant of the efficiency of job performance. However, blacks obtain lower average IQ scores than whites, so proportionately more blacks were rejected for jobs through being screened out as unsuitable on the grounds of failing to meet the required IQ. This became known as "adverse impact."

In 1970 the Equal Employment Opportunity Commission (EEOC) issued guidelines which defined job selection tests as discriminatory if they had an adverse impact on hiring blacks unless the tests could meet these extremely strict standards. In 1971 this recommendation was tested in the U.S. Supreme Court in the case of *Griggs v. Duke Power Company*. The Supreme Court supported the EEOC's recommendation and effectively made the use of intelligence tests for job selection illegal. Nevertheless, research continued to demonstrate that intelligence tests were useful predictors of job performance.

On the basis of this research, in 1981 the U.S. Labor Department's Employment Service issued a recommendation encouraging state employment services to use its General Aptitude Test Battery (GATB) more efficiently in their personnel selection procedures. The new procedure in fact aggravated the problem of the adverse impact of the use of the test against blacks. To overcome this problem, the Labor Department recommended the adoption of "race norming." This involved dividing the American population into blacks, Hispanics, and others (largely whites) and producing separate norms for each group. The effect of this policy was that, as Blits and Gottfredson put it:

> this sleight-of-hand gives bonus points to members of groups that score on average lower than other groups.[4]

By 1990, forty states were using race norms in their selection procedures.

Jan Blits, Ph.D.
University of Delaware

In 1986, the U.S. Justice Department challenged race norming in job selection on the ground that it constituted unconstitutional racial discrimination against whites and was therefore illegal. The Labor Department responded to this challenge by agreeing not to extend the use of race norming further until a panel of the National Academy of Sciences had made an evaluation of the procedure and made a recommendation. The National Academy of Sciences panel issued its report in 1988. This accepted that intelligence tests predict job performance but nevertheless recommended that the race norming practice should be continued.

This conclusion was criticized by Gottfredson in a paper written in collaboration with her colleague Jan Blits in 1990.[5] They argued that the panel was activated by political considerations, that its logic in supporting race norming was flawed, and that the continued use of race norming would have the effect of perpetuating racial inequality. The outcome of this dispute was that race norming was banned by the Civil Rights Act of 1991.

CONFRONTATION AT THE UNIVERSITY OF DELAWARE

In 1986 Linda Gottfredson left Johns Hopkins to take up an appointment in the department of educational studies at the University of Delaware. Her work at Delaware was supported by the Pioneer Fund from 1986 onwards. In 1990 her articles on intelligence and race norming attracted a certain amount of publicity and adverse criticism at the University. A faculty member urged the president to refuse to accept any further grants from the Pioneer Fund. The University administration began to harass Gottfredson in a number of ways. A committee investigated the content of her research, attempts were made to interfere with her teaching, her papers on race norming were classified as non-research for the purpose of merit review, and she was rejected for promotion to full professor. The University authorities also banned any further grants from the Pioneer Fund on the grounds that minority students might feel threatened by them.

Gottfredson contested these actions and eventually won favorable decisions from independent panels evaluating her case. A faculty senate committee concluded that Gottfredson's promotion committee had been unfair in its evaluation of her application for promotion and that the sociology department had violated her academic freedom when, for ideological reasons, it began denying sociology majors credit for her course in the sociology of education. The local chapter of the American Association of Professors criticized Gottfredson's dean for threatening her academic freedom and questioning the content of her teaching. The University's prohibition on accepting grants from the Pioneer Fund was reversed by a national arbitrator, who ruled that the University had acted improperly in examining and criticizing the content of Gottfredson's research.

The confrontation between the University and Gottfredson lasted almost three years. It was concluded in April 1992 with a victory for Gottfredson in settlement of which she was once more permitted to receive grants from the Pioneer Fund and was given a year's paid leave of absence. Later in 1992 the American Association of University Professors issued a statement based on the Gottfredson case, condemning any attempt by universities to deny a faculty member the opportunity to receive funding for ideological reasons, on the grounds that this is an infringement of academic freedom. In 1994, the University of Delaware endorsed the AAUP statement protecting external funding from political interference. Thus, what began as an attempt to isolate researchers from the Pioneer Fund ended in stronger protection for academic freedom.

THE EGALITARIAN FICTION

Following her vindication in her battle with the administration of the University of Delaware, Gottfredson has continued to carry out and publish research on the effects of racial differences in intelligence. She has examined how current programs to manage work force diversity are often counter-productive efforts to cope with unacknowledged racial disparities in skills and abilities in the workplace.[6] She also examined the social process by which social scientists collectively sustain and enforce what she calls "the egalitarian fiction" of racial equality. She writes that:

> social science today condones and perpetuates a
> great falsehood ... this "egalitarian fiction" holds

that racial-ethnic groups never differ in average developed intelligence .[7]

Gottfredson argues that many social scientists privately acknowledge that there are average racial differences in intelligence but that they are unwilling to go public on this because of a combination of money, politics, and fear. The result has been a triumph of intellectual dishonesty in which "ideology is declared knowledge and knowledge is dismissed as mere ideology".[8]

In a further article Gottfredson provided an overview of the status of research on intelligence and described what she called the "democratic paradox," namely, that two of the goals of democracy -- greater equality of opportunity and technological progress -- do not eliminate social hierarchies, but merely shift their basis from social advantage toward genetic advantage.[9] She believes that the way a society reacts to information about differences in intelligence is at least as important for its collective welfare as the differences themselves. She has delved further into this issue by analyzing how personnel psychology has become politicized.[10]

In 1996 the editor of *Intelligence* invited Gottfredson to produce a special issue, "Intelligence and Social Policy." This was intended to help to bridge the chasm between research on intelligence and social policy analysis, both by encouraging more experts on intelligence to take a broader, society-wide view of the meaning of intelligence and by encouraging more social policy analysts to become acquainted with the scientific evidence concerning intelligence. Gottfredson's own contribution to the volume was meant to counter the prevalent misconception that intelligence is not important in nonacademic settings and thus need not be taken seriously for most policy purposes. In particular, her paper concretely illustrated how higher or lower IQ levels confer systematic advantages or disadvantages in everyday affairs. These advantages or disadvantages, even when small, cumulate to affect individuals' overall life chances, especially in an increasingly complex, technological society.[11] Like much of her other work on the sociology of intelligence, this contribution showed, in addition, how the distribution of intelligence in a population shapes that society's limits and possibilities, choices and dilemmas.

Gottfredson is agnostic on the question of whether the black-white difference in intelligence has any genetic basis. In 1994 she wrote that:

> It is not clear yet why the disparities among groups are so stubborn -- the reasons could be environmental, genetic, or a combination of both -- but so far they have resisted attempts to narrow them.[12]

Whatever the factors responsible, she asserts that the intelligence difference between blacks and whites is an important cause of the differences in educational and occupational achievement and job performance. She maintains that the many social scientists, journalists and politicians who deny these differences are guilty of "the suppression of science," "collective fraud," "official mendacity," and "a comfortable lie."[13]

NOTES

1. Gottfredson, L .S. 1981. Circumscription and compromise: A developmental theory of occupational aspirations. *Journal of Counseling Psychology* (Monograph). 28: 545-579.

2. Gottfredson, L. S. 1985. Education as a valid but fallible signal of worker quality: Reorienting an old debate about the functional basis of the occupational hierarchy. In Kerchoff, A. (ed.) *Research in Sociology of Education and Socialization.* Vol. 5. Greenwich, CT: JAI Press.

3. Gottfredson, L. S. 1986. The *g* factor in employment. *Journal of Vocational Behavior.* 29: 293-296.

4. Blits, J. & Gottfredson, L. S. 1990. Equality or lasting inequality? *Society.* 27: 4.

5. *Ibid.*

6. Gottfredson, L. S. 1994. Dilemmas in developing diversity programs. In S. E. Jackson (ed.) *Diversity in the Workplace: Human Resources Initiatives.* New York, NY: Guilford.

7. Gottfredson, L. S. 1994. Egalitarian fiction and collective fraud. *Society.* 31: 53.

8. *Ibid.* 58.

9. Gottfredson, L. S. 1996. What do we know about intelligence? *American Scholar.* Winter: 15-30.

10. Gottfredson, L. S. 1994. The science and politics of race-norming. *American Psychologist.* 955-963.

11. Gottfredson, L. S. 1997. Why *g* matters: The complexity of everyday life. *Intelligence.* 24: 79-132.

12. Gottfredson, L. S. 1994. Egalitarian fiction and collective fraud. *Society.* 31: 54.

13. *Ibid.* 55-56.

Ralph Scott, Ph.D.
University of Northern Iowa

Chapter 39

Ralph Scott

Ralph Scott (born 1927) has been in the forefront of exposing the shortcomings of liberal innovations in American school systems designed to promote racially integrated schools by forced busing and mandating mixed ability classes. He has argued and produced evidence over a quarter of a century that these measures have no beneficial effects and are counter-productive for students of all races and abilities. The major factors determining educational attainment, Scott has argued, lie in prenatal, perinatal and family background, and family influences, rather than in schools.

Ralph Scott was born in 1927 and raised in a small Wisconsin town north of Madison, the son of a father who was an independent grocer-cum-butcher and a mother who was an elementary school teacher. In 1945 Scott was drafted into the U.S. Army and served in the staff office of the commander of the First Army. After two years of service, he entered Luther College in Decorah, Iowa, where he studied social science. Obtaining his baccalaureate degree, he entered the University of Wisconsin where he was awarded a master's degree in psychiatric social work. During the following decade, Scott worked as a school social worker and teacher in various inner-city and suburban schools. He became disillusioned, however, with the prevailing narrow egalitarian theories and programs which assumed that enrichment activities or counseling within schools could alone enable students to overcome significant learning problems.

In 1960 Scott entered the graduate school at the University of Chicago where he received his Ph.D. in 1964 in school and educational psychology. As instructor and clinical psychologist, he then taught at Northwestern University Medical School and collaborated with physicians in an appraisal of the genetic, physical, emotional, and developmental factors affecting mental and physical health. In 1965 Scott accepted a position as director of the Educational Clinic at the University of Northern Iowa, the leading training center for prospective teachers in Iowa. He became a licensed psychologist, a diplomate of the American Board of Professional Psychology and a fellow of the Society for Personality Assessment.

At the University of Northern Iowa, Scott again encountered the equalitarian ideology which virtually ignored genetic, prenatal, and developmental factors. It was generally assumed that learning problems could be corrected, or at least significantly ameliorated, solely through group curricular strategies within schools. In an attempt to correct this deficiency in educational strategies, he developed several published instructional materials which were designed to individualize children's learning needs through parental involvement.[1]

HOME START

By 1968 Scott was confident that schools alone could not significantly solve the learning problems faced by individual children, and he proposed a novel model for preschool education. This came to be known as "Home Start." Home Start was a program designed to serve families of black and white preschool children who resided in the low-income residential areas of Waterloo, Iowa. Home Start assigned primary teaching responsibilities to the children's parents. Concurrently, parents were provided psychological and medical assistance if their children required it. As a program, Home Start assumed that appraisal of children's cognitive potential was essential, and achievement and IQ tests were periodically administered to all participating children. Home Start thus contrasted radically in its philosophy with Head Start, the more widely known government program which assumed that all children would derive long term benefits if taken from their homes, provided with enrichment in a teaching center and then returned to mainstream. The Head Start model of the 1960s gave little direct responsibility to parents and assigned the task of teaching disadvantaged children to society. More recently there has been an increasing realization that children benefit from the assistance and encouragement of role model figures whom they respect, a role best filled by a concerned parent.

Ralph Scott's Home Start was designated the "Iowa Project" in 1970 by President Nixon's National Advisory Counsel on supplementary Educational Centers and Services. The program was designated the 1971 model for accountability for the state of Iowa. The President's Council then cited Home Start as one of 11 national projects to be employed throughout the nation as a model for other school districts. That same year, Home Start was selected as one of two national preschool models by the U.S. Department of Education and in 1975 was affiliated with the U.S. Department of Education National Diffusion Network.

For approximately a decade, Scott and his colleagues assisted school districts in over 40 states to develop the Home Start model.[2] By the mid-1970s, Scott collated and interpreted the results of Home Start and concluded that young children perform more effectively when parents consistently discharge their parental roles and when appropriate supplementary support services are provided in the case of physically and mentally handicapped children.

BLACK-WHITE DIFFERENCES

During his work on improving children's cognitive skills through Home Start, Scott noticed that when the effectiveness of the programs was evaluated, the IQs and educational achievement of black children fell significantly below that of white. The race difference was approximately one standard deviation or, in the case of intelligence, 15 IQ points. This difference was found to be present among two year olds and persisted through grade school and high school.[3]

From 1970 onwards Scott began to examine the national American evidence on black-white differences, working in collaboration with Jon Ford, Herbert Walberg, and other colleagues. They found that when blacks and whites were matched for educational background, blacks continued to obtain average test scores significantly below those of whites. The conclusion to be drawn was that inferior education could not explain the cognitive and educational disparities between blacks and whites.[4]

SCHOOL DESEGREGATION AND BUSING

In the early 1970s a number of social scientists began to advocate mandated school desegregation as a means of improving the educational achievements of black children. The thinking was that schools that were predominantly or exclusively black generally provided poorer educational facilities than those attended by whites. Furthermore, the segregation of blacks and whites in their own schools was believed to foster racial antagonism. If blacks and whites were educated together, the theory was, they would form friendships across the racial divide and come to like each other. Thus racially integrated schooling would achieve the dual purpose of promoting both the educational attainments of blacks and racial harmony.

The reason that many schools were segregated by race was that blacks and whites typically lived in different neighborhoods and their children attended neighborhood schools. This problem was overcome by sending black children to previously all-white schools, or alternatively by sending white children to previously all-black schools. To achieve these racially integrated schools, children were to be transported by bus, and the policy became known as "busing."

Busing programs were implemented in a number of states and cities during the 1970s, including Scott's state of Iowa. Scott argued the case against busing on the grounds that the cognitive differences between blacks and whites were not caused by inequalities between schools, since they were present in preschool children and also in historically integrated schools which had served black and white students living in the same neighborhoods. On these grounds he argued that racially integrated schools would not bring any benefit to either black or white children. Furthermore, he argued that busing programs would lead to a white flight from neighborhoods to which black children were bused, to breakdown of neighborhood communities, and to an increase in school violence.[5]

In 1976 and 1977 Scott obtained grants from the Pioneer Fund to the University of Northern Iowa to organize four public symposia in major cities with balanced panels of experts to discuss the educational issues involved in busing. Experts representing various disciplines advanced arguments for and against busing and drew significant support from minority parents and educators. One of the results that emerged from these symposia was that many blacks were opposed to busing. For instance, an opinion poll carried out in Boston in 1978 showed that 80 percent of black parents preferred their own neighborhood schools.[6]

By the early 1980s it had become clear that, while many experts asserted that busing produced black achievement gains, it had created significant problems in many schools. In 1982 the National Institute of Education appointed Scott as an academic advisor to its desegregation panel. At Scott's suggestion, funds were allocated to investigate further the question of whether black students learn more effectively in desegregated schools. The design Scott developed, with the assistance of NIE officials, called for seven of the nation's experts to identify the most significant studies into the busing-black achievement thesis and then to subject the data to statistical analysis.

This proposal was adopted and the experts identified the 157 most significant studies on the topic. Of these only 19 were sufficiently strong to warrant serious consideration. None of those studies provided any evidence that busing improved black achievement. Conversely, the literature indicated that busing was harmful to black students and caused interracial conflict. Despite the absence of evidence that busing improved black attainment, several scholars insisted that, with more data, the benefits of busing would be apparent.

With Scott serving as panel moderator, the seven experts presented their findings at the National Press Club on 17 September 1982. When several experts publicly stated that busing significantly enhanced black achievement, Scott asked for evidence, and the panel, including members who claimed busing achieved benefits, concurred there was no statistical evidence of long term significant educational gains derived from busing. Furthermore, it was admitted that some studies indicated that forced busing was harmful to black students. Several media representatives and several senior officials of the U.S. Department of Education asked Scott whether the obtained results, if published, might not retard desegregation. Scott replied that if forced busing was harmful, then black children should be entitled to attend neighborhood schools.

Scott next received an invitation from Thomas Ascik, the Associate Director of the National Institute of Education (the research division of the U.S. Department of Education), to draft a monograph on the panel's conclusions regarding the effects of school busing on black achievement. Scott wrote the monograph setting out the panel's conclusions on busing's negative effects and sent it in, but the National Institute of Education (NIE) ignored it. On 18 September 1983, Scott called Oscar Uribe, the NIE official responsible for the publication of the monograph, to ask when it was going to be published. Uribe told Scott that the monograph had been shelved and that Scott's contract as an NIE consultant was terminated. Evidently Scott's conclusions were not politically acceptable at NIE. The next year the NIE published its own monograph on busing without any mention of Scott's views or of his name. This anonymous document asserted that "there is a real possibility that busing significantly enhances the learning and educational attainment of black students." Nevertheless, Scott published independently seven papers over the years 1982-86 showing that desegregating schools and busing had no beneficial effects on black school children's educational attainments.[7]

TRACKING AND MIXED ABILITY CLASSES

By the mid-1980s, American political leaders were expressing increased concern over the educational standards of American schools.[8] International comparisons generally placed American students at or near the bottom of international league tables, when comparisons were made with students of other industrialized nations.[9] Pursuing these issues, Scott wrote two commissioned papers for the U.S. Department of Education. In the first of these papers Scott focused on the achievement opportunities provided for gifted American students and concluded that America's most capable young people were being academically under-challenged. One important reason for this was that many of America's most talented students were being placed in "untracked," or mixed ability classes, not on the basis of their readiness and ability to learn material, but merely to assure an appropriate racial mix. Scott concluded that mixed ability classes reduced the quality of instruction for talented students and were harmful to students of all abilities.[10]

In his second paper Scott examined additional factors within American schools which facilitate or impair students' educational attainment. From an examination of the research evidence, Scott concluded that spending had little effect on the quality of education and that family background factors were the major determinant of student learning. He suggested that schools place greater emphasis on strengthening the bonds between home, school, and community, and he urged that schools formulate and implement policies designed to focus on the instructional needs of individual students.[11]

In 1985 Scott was appointed chairman of the Iowa Advisory Commission to the U.S. Commission on Civil Rights. During his tenure on this Commission, Scott encouraged the Commission to examine the effects of desegregation on students, families, and communities. He also urged scientific investigations into the impact of imposing ethnic quotas with respect to such issues as school discipline and suspensions, and placements in classrooms for the mentally retarded or gifted. However, these recommendations were strongly attacked in certain sections of the media by those who favored ethnic quotas and insisted on classifying students by race rather than by ability or performance. These attacks proved persistent and, since Scott had no wish to abandon his scholastic research in order to engage in public disputes, he resigned as Iowa chair in 1989, despite the support of the director of the U.S. Commission on Civil Rights.

In the 1990s Scott has continued to question the widespread replacement of tracking by mixed ability classes and to attack the continued use of busing as an ideologically-motivated egalitarian attempt at social engineering which the research evidence has consistently shown has no beneficial effects for either black or white children.[12]

NOTES

1. Scott, R. & Dunbar, A. 1968. *LRS Seriation Test.* New York: Harper and Row; Scott, R. 1968. *Learning Readiness System: Classification and Seriation.* New York: Harper and Row.

2. Scott, R. 1970. *Home Start II: Individualized Preschool Enrichment for Vulnerable Children, Title III ES.* Washington D.C.: U.S. Department of Education, National Diffusion Network.

3. Kobes, D. & Scott, R. 1975. The influence of family size on learning readiness profiles of socio-economically disadvantaged preschool blacks. *Journal of Clinical Psychology.* 31: 85-88; Scott, R. 1976. Home Start: third grade follow-up assessment of a family-centered preschool program. *Psychology in the Schools.* 13: 435-438; Scott, R. 1986b. Mandated school busing and student learning: achievement profiles of third, fifth and tenth grade black and white students. *Mankind Quarterly.* 27: 45-62.

4. Ford, J. & Scott, R. 1971. *An Assessment of the Differences between High and Low Achieving Students.* Washington, D.C.: National Center for Educational Research and Development; Scott, R. & Walberg, H. 1979. Schools alone are insufficient. *Educational Leadership.* 45: 24-27.

5. Scott, R. 1981. *Mandated Desegregation: Overview of the Literature.* Position paper prepared for the Attorney General, State of Missouri, 12 December.

6. Scott, R. 1977. *Harmfulness of Enforced Busing: an Emerging Hypothesis.* Boston: Fourth Symposium on Constructive Alternatives to Forced Busing. Boston Globe, January 17.

7. Scott, R. 1986. Busing — the remedy that failed. *Journal of Social, Political and Economic Studies.* 11: 189-200.

8. Scott, R. 1988. Education and ethnicity: the US experiment in school integration. *Journal of Social, Political and Economic Studies.* Monograph 17.

9. Lynn, R. 1988. *Educational Achievement in Japan.* London and New York: Macmillan.

10. Scott, R. 1984. *Achievement Profiles of Gifted and Talented Students in Secondary Schools.* Washington, D.C.: U.S. Department of Education.

11. Scott, R. 1984. *Productive Factors Which Influence Levels of Learning.* Washington, D.C.: U.S. Department of Education.

12. Scott, R. 1993. Racial quotas in public school classrooms. *Mankind Quarterly.* 33: 309-320; Scott, R. 1993. Untracking advocates make incredible claims. *Educational Leadership.* 51: 79-81.

Garrett Hardin, Ph.D.
University of California at Santa Barbara

Chapter 40

Garrett Hardin

Garrett Hardin (born 1915) has been the leading advocate over the last three decades of the 20th century for the necessity of controlling the growth of world population and of reducing immigration into the United States. He has written numerous books and articles to promote these objectives. His work has consisted of four principal themes. The first is that the world is experiencing a population explosion which will deplete natural resources, damage the environment, and reduce the quality of life. To prevent this from happening, ways need to be found to reduce world population growth. Second, while it is going to be difficult to stabilize population growth in the economically developing world, it can be stabilized in the United States by a reduction of immigration. Third, he believes that multicultural societies are inevitably fraught with social division and conflict and this is another reason why immigration into the United States should be reduced. Fourth, the aim of reducing or stabilizing population growth needs to be supplemented by the principle that it would be desirable to control not only the quantity of children but also their quality.

Garrett Hardin was born in Dallas, Texas in 1915. His father was a freight sales representative with the Illinois Central Railroad. Although the family moved frequently because of his father's job, they had secure roots in his grandfather's farm in southwestern Missouri. Hardin's high school and college days were spent in Chicago. He showed promise in writing from an early age. At the age of 15 he won a city-wide contest run by the *Chicago Daily News* with an essay on the importance of Thomas Edison. For this he was awarded a trip east to visit the aging inventor.

In 1932 Hardin won both a University of Chicago academic scholarship and a dramatic arts scholarship at the Chicago College of Music. A month's attendance convinced him that he could not follow both paths simultaneously, and so he abandoned the dramatic scholarship. In 1936 Hardin graduated from the University of Chicago in zoology, studying under the ecologist W. C. Allee. He then transferred to Stanford University, where he obtained his Ph.D. in microbial ecology in 1941. His most influential mentors were the microbiologist C. B. van Niel and the geneticist George W. Beadle, later to be awarded the Nobel prize. Shortly after graduation Hardin began work at the Carnegie Institution of Washington's Division of Plant Biology, which had a laboratory on the Stanford campus. For four years he was part of a team investigating antibiotics produced by algae, as well as the future possibility of using cultured algae as animal food.

In 1946 Hardin resigned his research position at the Carnegie Institution to accept an associate professorship at the University of California's campus in Santa Barbara. During the next two decades he devoted much of his time to developing an ecologically-oriented course in biology for the general citizen, which he adapted to closed-circuit television. He was appointed full professor of human ecology in 1963. Hardin's work on population control and immigration reduction has been supported by grants from the Pioneer Fund from 1988 through 1992.

THE TRAGEDY OF THE COMMONS

Hardin achieved a major impact for his views on the desirability of reducing the world population explosion in 1968 with his presidential address, "The Tragedy of the Commons," delivered to the Pacific Division of the American Association for the Advancement of Science.[1] Hardin began his lecture by reference to the point made by an English mathematician named William Lloyd in a pamphlet published in 1833. Lloyd argued that if a public purse were made available for everyone to dip into, the money in the purse would rapidly disappear. Hardin suggested the alternative analogy of a common land on which everyone is allowed to graze their cattle. When this right is available, there is a natural tendency for people to exploit the grazing to the full because the gain to them as individuals outweighs the cost. The result is that the common land becomes overgrazed and deteriorates. Hardin argued that this was inevitable and called it "the tragedy of the commons." He noted that this problem had been solved by the introduction of property rights in land. Once land became owned by individuals, rather than in common, it became in the owners' own interest not to overgraze it and to maintain its productive capacity.

The general principle of the tragedy of the commons is that individuals will exploit anything that is free in order to maximize their own advantage, but that this entails a cost to society as a whole. Hardin then applied this principle to the production of children. People who have a large number of children, he argued, are imposing a cost on society, which they themselves do not have to bear. "Freedom to breed," he wrote, "will bring ruin to all."[2] How, therefore, can we prevent people from damaging the public well-being by producing excessive numbers of children? Hardin observed that those concerned with this problem were making appeals to the conscience of the offenders. He argued that this would not be effective, partly because it would not work and partly because it would generate guilt, and that Freudian psychoanalysts have demonstrated that guilt is psychologically damaging.

Since appeals to the conscience of the group he called "the nations' (or the world's) breeders" would be both ineffective and psychologically undesirable, Hardin argued that coercion would be necessary to prevent people from having excessive numbers of children. He recalled that the United Nations issued a statement in 1967 to the effect that it was a natural right of couples to have as many children as they wished, but he said that this had to be rejected. He recognized that the restriction of people's right to have unlimited numbers of children would necessarily involve a reduction in individual freedom. Nevertheless, this was justifiable for the good of society as a whole, just as the freedom to rob banks is curtailed by the criminal law. Hardin concluded:

> The only way we can preserve and nurture other and more precious freedoms is by relinquishing the freedom to breed, and that very soon .[3]

Neither in this address nor subsequently has Hardin suggested the ways by which people's right to have children would actually be curtailed. Presumably they would be punished in some way if they exceeded the permitted limit, or possibly they would be compulsorily sterilized. Hardin leaves his readers to work these details out for themselves. But although the measures for reducing birth rates are not spelled out, Hardin made it clear that some kind of sanctions would be required to enforce family limitation. Hardin has reiterated and elaborated the themes in his 1968 lecture on a number of occasions over the course of the succeeding quarter century.

POPULATION, RESOURCES, AND POLLUTION

Hardin's basic argument is that the earth has a limited carrying capacity for the size of the population it can accommodate. He believes that the optimum carrying capacity had been reached by the last quarter of the 20th century and any further increases in world population will bring about a deterioration in the quality of the environment and of human life. As the numbers of people increase, there will inevitably be rising levels of pollution, degeneration of the quality of agricultural land, deforestation, and deterioration of air and water quality. To prevent this deterioration, Hardin believes our first aim should be to arrest the growth of world population.

Hardin notes that fertility in the United States and Europe fell to about two per woman or even lower in the 1980s and that this would stabilize the size of the populations at approximately their present numbers for several decades to come. He welcomes this development. To ensure that fertility remains low Hardin advocates a variety of measures including subsidized birth control and abortion, paying adolescent girls an annual allowance conditional on their not having a child, the abolition or reduction of tax allowances for children to discourage people from having them, and rewards for those who have only one child or none, which might take the form of prestigious subsidized vacations.[4]

IMMIGRATION

Because fertility is low in the United States and Europe, Hardin believes that the problem of excessive population growth would be largely under control if it were not for immigration. He sees immigration as the major problem that will lead to increases in population in the economically developed world. To prevent this growth he advocates the reduction of immigration nearly to zero. In a striking metaphor, Hardin has on several occasions used the analogy of a nation as a lifeboat.[5] A lifeboat can only hold a certain number of people. If more are taken on board, the lifeboat sinks and everyone will be drowned. The only rational course of action for those in a full lifeboat is to refuse to take anyone else on board. It is the same with a nation. "To survive," he writes in his last book, *Living Within Limits*:

> rich nations must refuse immigration to people who are poor because their governments are unable or unwilling to stop population growth. [6]

Two years later he reaffirmed this reasoning in a journal article, this time drawing an analogy from microbiology.[7] Biologists see immigration as a developmental phenomenon. ... Just as the thyroid gland withers away during growth from babyhood to adulthood, so too must immigration disappear as the country matures by becoming filled up.[8]

Multiculturalism

Hardin advances another reason for reducing immigration. This is that most immigrants into the United States following the 1965 Immigration Act have been Mexicans, blacks from the Caribbean and Africa, and Asians. This is bringing about an increasingly multicultural society, and Hardin believes that this is a recipe for social disorder. He writes:

> Diversity within a nation destroys unity and leads to civil wars. Immigration, a benefit during the youth of a nation, can act as a disease in its mature state. Too much internal diversity in large nations has led to violence and disintegration.[9]

In 1991 he wrote that the cult of multiculturalism has been responsible for the large scale immigration of non-European peoples into the United States and this will destroy social unity:

> We are now in the process of destabilizing our own country through the unlimited acceptance of massive immigration. The magic words of the destabilizers are "diversity" and "multiculturalism." Diversity is good, yes: but like all good things, it is possible to have too much of it *in one place*. The telling example of our time is Beirut. For a while the diversity of this city was beautiful and exciting, it was called the Paris of the Mediterranean by the Arab millionaires who flocked to it. But as it grew in population, and as the proportions of the disparate ethnic groups changed, peace vanished. Within the bounds of a single nation the mutual stresses of intolerant groups became too great.

Popular anthropology came along with its dogma that all cultures are equally good, equally valuable. To say otherwise was to be narrow-minded and prejudiced, to be guilty of the sin of ethnocentrism. In time, a sort of Marxist Hegelian dialectic took charge of our thinking: ethnocentrism was replaced by what we can only call ethnofugalism -- a romantic flight away from our own culture. That which was foreign and strange, particularly if persecuted, became the ideal. Black became beautiful, and prolonged bilingual education replaced naturalization.... Idealistic religious groups, claiming loyalty to a higher power than the nation, openly shielded and transported illegal immigrants.

If two cultures compete for the same bit of turf (environment), and if one of the populations increases faster than the other, then year by year the population that is reproducing faster will increasingly outnumber the slower one. If, "other things being equal," there are advantages to being numerous, then in time the slowly reproducing population will be displaced by the fast one. This is passive genocide. It may be that no one is ever killed, but the genes of one group replace the genes of the other. That's genocide.[10]

THE THIRD WORLD POPULATION EXPLOSION

While the size of the population in the economically developed world has approximately stabilized in the last quarter of the 20th century except for immigration, population growth remains high in the economically developing world of Latin America, Africa and much of Asia south of the Himalayas. Hardin believes that ways need to be found for halting this excessive increase. He does not accept the theory of many demographers that as people become more affluent they automatically control their fertility, and its implication that the economically developed nations should give more aid to the underdeveloped nations to bring about the required increase in affluence.

Hardin holds the contrary Malthusian view that economic and other forms of aid simply lead to more babies being born and surviving. Aid increases the size of the populations of third world countries so that they will need yet more aid in the future. For instance, Hardin states:

> sending food to Ethiopia does more harm than good. Each year the production from Ethiopian land declines. The lands are used beyond their carrying capacity because there are far more people than renewable resources. [11]

Hardin's prescription for this problem is for the first world nations to cease to give aid to third world countries and let them solve their own problems of adjusting their population size to the productive capacity of their lands. The only aid that the United States and other rich countries should give to the impoverished third world is information about birth control and contraceptives. Hardin is aware that some people will call the denial of aid to starving third world populations genocide, but he regards his prescription as being to the long term advantage of the third world countries. He writes:

If a country is poor and powerless because it already
has too many children for its resources, it will
become even poorer and more powerless if it breeds
more.[12]

Hardin regards the desire of many people in
the United States and Europe to send aid to third
world countries as what he calls "promiscuous
altruism" and "short range compassion." Some of
these people, he believes, are what he calls
"ethnofugalists" who see virtue only in others and
are the opposite of ethnocentrists who see virtue
only in their own ethnic group.[13]

Hardin does not offer any detailed advice to
governments of third world countries on how to
control the growth of their burgeoning populations,
but he is clear that some form of compulsion will
probably be necessary and is justified. "Like it or
not, the issue of coercion must be faced" he writes,
and continues "the present generation has become
pathologically sensitive to the word *'coercion'*."[14]
He writes with approval of the Chinese population
control policy in which women members of
production teams have to seek permission to have
a child. If they become pregnant after permission
has been refused, they are required to have an
abortion. Hardin gives his approval to the Chinese
policy of allowing couples to have only one child
and the imposition of financial sanctions on those
who disregard it. His only regret is that the one
child policy has not been working in rural China.[15]
Evidently, the punishments for having more than
one child have not been fully effective and, by
implication, needed to be strengthened. He is

critical of the American government for cutting off aid to China for the promotion of birth control when the widespread use of abortion became known.

POPULATION QUALITY

Although Hardin's principal concern has been the growth of population numbers, he has also voiced concern about population quality and it is here that his writings have a eugenic dimension. He has criticized Paul Erlich's 1967 book *The Population Time Bomb* and the American organization Zero Population Growth (ZPG) on the grounds that Erlich and ZPG failed to take into account the issue of population quality. Erlich argued that the world population explosion was so serious that people in the United States and Europe had a duty not to have children as a contribution to reducing world population. Hardin argues that this would be dysgenic because the peoples of the first world are more intelligent than those in the third world. The proper solution to the world population explosion, Hardin argues, is for each country to stabilize its own population numbers.

Similarly, Hardin criticizes the Zero Population Growth movement because its message of the desirability of reducing the birth rate appeals largely to college graduates. If college graduates respond by having fewer children but non-graduates do not, the result will be dysgenic. Hardin says that:

To put it bluntly it would be better to encourage the breeding of more intelligent people rather than the less intelligent. ZPG's entire attraction has been among the college population. So, in effect, ZPG is encouraging college-educated people to have fewer children instead of encouraging reduced fertility among the less intelligent.[16]

Hardin is aware that many economists dispute his claim that world population has already reached its optimum, but he castigates them for their failure to recognize the principles of limited natural resources, diseconomies of scale and the complexity of ecological systems which are easily destroyed by human exploitation. The thinking of economists who fail to recognize these principles, he asserts, is distorted by the Freudian process of denial, by which uncongenial realities are repressed into the unconscious mind.[17]

Garrett Hardin has received many honors. In 1973 he was elected to the American Academy of Arts and Sciences and in 1974 to the American Philosophical Society. In 1979 he was awarded the Margaret Sanger Award for his support for the wider provision of birth control and population limitation. In 1993 he was one of the recipients of the Phi Beta Kappa annual book prizes at which the chairman of the award committee described Hardin's *Living within Limits* as:

a trenchant, learned, passionate analysis of the most difficult problem that confronts mankind since the threat of nuclear annihilation has dwindled -- the threat of an apparently inevitable human over-population of the earth.[18]

NOTES

1. Hardin, G. 1968. The tragedy of the commons. *Science*. 162: 1243-1248.

2. *Ibid.* 1248.

3. *Ibid.* 1248.

4. Hardin, G. 1993. *Living within Limits*. New York, NY: Oxford Univ. Press.

5. Hardin, G. 1974. Living on a lifeboat. *Bioscience*. 24: 561-568.

6. Hardin, G. 1993. *Op. cit.* 294.

7. Hardin, G. 1995. Multiculturalism: a recipe for conflict. *Public Affairs*. 5: 42.

8. Hardin, G. 1993. *Op. cit.* 42.

9. *Ibid.* 42.

10. Hardin, G. 1991. Conspicuous benevolence and the population bomb. *Chronicles*. 15: 20-22.

11. Hardin, G. 1993. *Op. cit.* 37.

12. *Ibid.* 252.

13. *Ibid.* 297.

14. *Ibid.* 270.

15. Spencer, C. 1992. Interview with Garrett Hardin. *Omni*. 14: 55-63.

16. *Ibid.* 59.

17. Hardin, G. 1993. *Op. cit.* 274.

18. *The Key Reporter*. 1993/1994 Winter: 1.

Afterword

The Achievements of the Pioneer Fund

This book has recounted the first 60 plus years of the Pioneer Fund, nearly the only nonprofit foundation making grants for study and research into individual and group differences, and the hereditary basis of human nature. In addition to listing many of its accomplishments and examining some of the controversies in which it has been involved, it has provided thumbnail biographies of the scientists, scholars, and public figures associated with it. Over those 60 years, the research funded by Pioneer has helped change the face of social science.

Since its inception, the Pioneer Fund has sought out excellence in both its directors and its grant recipients. The list of Pioneer directors has included a U.S. Supreme Court Justice, a decorated army general who was a member of the UN Atomic Energy Commission, a president of the American

Psychological Association, and a Nuremberg tribunal prosecutor.

Pioneer Fund researchers have been equally distinguished. At the time of this writing, two are among the five recent psychologists most cited by other scientists. Pioneer grantees have been elected as the presidents of the American Psychological Association, the British Psychological Society, the Behavior Genetics Association, the Psychonomic Society, the Society for Psychophysiological Research, the Psychometric Society, and the National Council on Measurement of Education. One grantee won a Nobel prize, two were Guggenheim Fellows (one for doing Pioneer-funded work), and three more were selected by the Galton Society of the United Kingdom to give the 1983, 1995, and 1999 annual Galton Lectures, and one was selected to give the quadrennial Spearman Lecture (also on the basis of Pioneer-funded work). Three are among the eleven recent biographees in the *Encyclopedia of Human Intelligence*, and ten of the articles in that two volume work were written by grantees. At the time of this writing, Pioneer grantees serve on the editorial boards of major academic journals, including three on the board of *Personality and Individual Differences*, and three more on the editorial board of the journal *Intelligence*, and two have served on the editorial board of *Behavior Genetics*, and many have published in these journals.

Most of the Pioneer grantees hold Fellow status in one or more scientific organizations. Many have won academic honors from the American

Association for the Advancement of Science, the American Educational Research Association, the Center for Advanced Study in the Behavioral Sciences, Mensa, Educational Testing Service, and the American Psychological Association. Pioneer funded some of the research for which the honors were awarded. Thirty recent Pioneer grantees have together published close to 200 scholarly books and 2,000 scientific articles, most in the leading peer-reviewed journals.

CHANGING THE FACE OF SOCIAL SCIENCE

Since its beginning in 1937 the Pioneer Fund has supported scientific research in four major areas: (1) the nature of intelligence, (2) behavioral and medical genetics, (3) race differences, and (4) IQ, population, and related social problems. The Fund has also supported the dissemination of information in all these research areas to the general public. Since several of these scientific areas were avoided for a long time by larger foundations as too controversial, Pioneer is proud of its initiative and success in helping to reshape the face of social science by supporting cutting edge research into once tabooed topics. By doing so, the Pioneer Fund has helped change the face of social science, re-establishing its Darwinian-Galtonian origins.

The evolutionary-genetic approach that posits both hereditary and environmental factors and examines their interaction has today become the mainstream view of professionals in behavior genetics and psychometrics (the testing and measurement of intelligence and personality). In part, the objective economic efficiency of psychological testing in education and job selection has helped win this intellectual battle. So too has the fact that the science section of any major Sunday newspaper usually contains a report of new evidence for the genetic basis of yet another human medical or behavioral trait. The opponents of the Darwinian-Galtonian tradition, meanwhile, have been forced into fighting rearguard actions that rely almost exclusively on anecdotal accounts devoid of pragmatic value at best and *ad hominem* and false claims attacks at worst.

Despite the progress on the behavioral genetic and psychometric fronts, intellectual conflict still rages regarding the study of IQ, population, and social policy, and, especially, race differences. A harbinger that the intellectual tide has started to turn even in these fields is the front page review of three books — Pioneer-supported Itzkoff's *The Decline of Intelligence in America*, Pioneer-supported Rushton's *Race, Evolution, and Behavior*, and Herrnstein and Murray's *The Bell Curve* (which made extensive reference to Pioneer-funded research) — that appeared in the *New York Times Book Review* of 16 October 1994. *Times* science writer Malcolm Browne concluded his favorable review by throwing down the challenge to critics of Pioneer-funded researchers that:

> the government or society that persists in sweeping their subject matter under the rug will do so at its peril.[1]

It is appropriate to conclude this work by highlighting the scientific research the Pioneer Fund has supported in each of these areas, the leading edge of the re-establishment of the Darwinian-Galtonian paradigm in the social and behavioral sciences.

THE NATURE OF INTELLIGENCE

Intelligence was initially formulated as a psychological concept, but it was quickly realized that it must have some neurophysiological basis and that this needed to be understood. Pioneer supported scholars have made a major contribution to this problem and have established two neurophysiological bases for intelligence. The first of these is the size of the brain; the second is the speed and accuracy of neural transmission. Heritability studies provide further evidence for a biological component to intelligence.

The Neurophysiological Basis of Intelligence. The early evidence for a positive relationship between brain size and intelligence was assembled by Donald Swan in the 1960s. In the 1980s and 1990s further evidence was collected by Jensen, Lynn, Rushton and P. A. Vernon. By the mid-1990s this neurophysiological component of intelligence had been well documented.

A second neurophysiological basis of intelligence, the speed and accuracy of neural transmission of information through the brain and nervous system, has been investigated by the late Eysenck, Jensen, Lynn, Reed, and P. A. Vernon.

Psychometricians have also studied the structure of intelligence, whether there is a single factor of general intelligence, a number of independent abilities, or a hierarchical structure with general mental ability at the topmost level. Most psychometric evidence supports the hierarchical model, of which Jensen is today's leading exponent.

The Heritability of Intelligence. A strong genetic component in the determination of intelligence has been shown in a series of studies on twins and adopted children from the late 1930s onwards. Pioneer scholars have not been the only people to do work on this issue, but they have made substantial contributions to it.

Osborne has shown that identical twins are more closely similar for intelligence than non-identicals, indicating the presence of genetic factors, and that this is true for both blacks and whites. Lynn has shown that intelligence has a high heritability among the Japanese. Bouchard and the Minnesota group have shown the similarity for intelligence of identical twins reared in different families, indicating a high heritability. Horn, together with Loehlin and other colleagues at the University of Texas, has studied adopted children and shown a resemblance between their intelligence and that of their natural mothers, again indicating a high heritability. Reed has adduced further evolutionary arguments for the heritability of intelligence. The consensus view based on the work of Pioneer scholars and others indicates that the heritability of intelligence among adults is around .80.

BEHAVIORAL AND MEDICAL GENETICS

Since Galton, social scientists have attempted to assess the respective roles of nature and nurture, genes and environment, in shaping not only our bodies, but our behaviors. Among the most powerful methods of modern human behavior genetics are comparisons of identical and fraternal twins, and of biological and adopted children. Pioneer scientists have conducted some of the most important studies using these methods. Some Pioneer research has also been in medical genetics, especially the early detection of genetic disorders.

The Heritability of Special Abilities. Some Pioneer scholars have tackled the question of whether specific abilities have some heritability over and above the heritability of general intelligence. Horn and his colleagues have shown that this is so for spatial and perceptual speed abilities. Thomas has argued the case that spatial ability is an inherited X-linked recessive gene and that this explains the higher average level of this ability in males.

The Heritability of Personality. Similar progress has been made in the understanding of the neurophysiology and heritability of personality. Lykken has shown that the personality disorder of psychopathic personality, which is responsible for most serious recidivist crime, is characterized by weak neurophysiological responses to stress, resulting in low levels of anxiety and poor social learning of the moral rules of society. He has also shown that this and other personality characteristics have a substantial heritability. Eysenck reached a similar conclusion. His lifetime of work was dedicated to showing that his three major personality traits of neuroticism, introversion-extraversion, and psychoticism all have heritabilities of around .50.

Eysenck also showed that crime has a heritability of approximately .60. Bouchard and his group have shown that the heritability of work motivation is about .50. Horn and his colleagues have demonstrated in their study of adopted children a high heritability for psychopathic personality. Rushton has examined the London twin register and found a high heritability for altruism. All of these have shown that different expressions of temperament and personality have substantial heritabilities.

Medical Genetics. In addition to their concerns about intelligence and temperament, geneticists and social scientists have also investigated genetic disorders. The Pioneer Fund has supported medical research on several genetic disorders including sickle-cell anemia, hemophilia, Tay-Sach's disease and schizophrenia. Pearson and Caton have also described how advances in the prenatal diagnosis of affected fetuses are reducing the birth incidence of some of these disorders.

RACE DIFFERENCES

In addition to the study of the part played by heredity in the differences in intelligence and personality between individuals, there is the problem of the contribution of heredity to differences between racial and ethnic groups. The principal research questions tackled by a number of Pioneer scholars have been the difference in average IQ of blacks and whites in the U.S. and the implications for applicant selection in schools and employment.

Race and IQ. In the early 1950s McGurk made an important contribution by showing that American blacks perform relatively better on intelligence tests of cultural knowledge than on tests of abstract reasoning. This was contrary to the cultural deprivation theory popular at the time which maintained that the low performance of blacks was due entirely to cultural deprivation and inferior schooling.

In the late 1950s and early 1960s Shuey reviewed virtually all of the published studies that had been carried out in the United States on black-white differences in intelligence. She found that the disparity had not narrowed over the period of nearly half a century, from the 1910s to the early 1960s. In an updated survey Osborne and McGurk found that the black-white difference was still present in the 1970s. Humphreys and Lynn have shown that it remained undiminished in the 1980s.

From 1969 onwards one of the leading exponents of the case for a genetic component in IQ differences has been Jensen. Much of his work has consisted of the systematic refutation of the socio-economic and cultural explanations for the IQ difference, such as that blacks are handicapped by the poor quality of their education, their low socioeconomic status, cultural bias in the tests, low motivation, poor nutrition and adverse prenatal and perinatal care. Jensen has shown that none of these is capable of explaining the black-white difference.

Several Pioneer scholars have carried out research and written on the issue and reached the same conclusion. Eysenck, Itzkoff, Levin, Lynn, Rushton, and P. E. Vernon all examined this problem. Jensen has also advanced the theoretical understanding of the nature of the black-white difference in intelligence by showing that it consists largely of a difference in Spearman's g, the general factor in all mental activities.

From the mid-1970s four more persuasive arguments for a genetic basis to race differences in intelligence were advanced by Pioneer scholars. First, there was the increasing evidence that the Asian peoples of the Pacific rim have slightly higher average IQs than North American and European Caucasians, which showed that IQ tests are not inherently biased in favor of whites. Second, the evidence reviewed by Swan in the 1960s that there are race differences in average brain size was greatly strengthened by new evidence produced by Rushton, which provided a neurophysiological basis for some of the racial differences in intelligence. Third, Lynn's and Levin's analyses of transracial adoption studies showed that black babies adopted by white parents did not register any lasting gains in intelligence. It had often been argued that if blacks were brought up in a white environment they would attain the same IQs as whites, but the transracial adoption studies showed that this was not so. Finally, Jensen has shown that two processes known to be biological, regression to the mean and inbreeding depression, can be analyzed so as to verify the genetic component of race differences.

All these new lines of evidence have important implications in education and employment.

Race Differences in Personality. Several Pioneer scholars have raised the question of whether there may be race differences in personality traits and characteristics, beyond the IQ factor. In the 1960s Kuttner was one of the first to document the high crime rates of blacks in the United States. In the 1970s and 1980s this phenomenon was analyzed by Gordon, who argued that the higher crime rate of blacks could be explained by their lower average IQ, since there is a general association between low test scores and crime.

However, other Pioneer scholars have proposed that, on average, groups differ in the personality characteristics that predispose an individual to crime. Levin has suggested that blacks are less given than whites to deferring gratification and weighing current acts against future consequences. Lynn has provided evidence that blacks have higher levels of testosterone, the male hormone contributing to aggressiveness and violence. The most comprehensive theory of race differences in personality characteristics is Rushton's *r-K* theory. It states that, of the three major races, Asians are the most biologically predisposed toward family stability and the least so toward crime, that blacks are the most predisposed toward higher levels of extraversion and aggression and less inclined toward family stability, with Caucasians being intermediate the two other groups.

IQ, POPULATION, AND SOCIAL POLICY

From the closing decades of the 19th century a number of social scientists have been concerned with changes in IQ in the economically developed nations. Some evidence indicates that it has been declining as a result of those with higher IQs having fewer children. Demographers refer to this as dysgenic fertility. There is also some evidence for increases in IQ, possibly because of better nutrition, among the lowest SES groups in these nations.

The broad variation in IQ is also a concern, especially given the global economy's increasing demand for cognitive and symbolic-analytical ability. Social policies that ignore differences in IQ and even personality traits such as conscientiousness are less likely to succeed and more likely to fail than those that recognize them. A number of scholars, including Pioneer grantees, have examined the complex matrix of relationships between IQ, education, employment, immigration, demographics, and social organization.

The Pioneer Fund's first project, directed by Flanagan, was research in positive eugenics consisting of the offer of financial assistance to U.S. Army Air Corps officers to have additional children. Seven children were born as a result of this plan. Clearly this did not significantly affect the American gene pool. The cost of this project was considerable and the principal lesson was that it would be feasible but also expensive to induce the professional classes to have more children by the provision of financial incentives.

Several other Pioneer scholars have been concerned with problems of population and social policy. Shockley drew attention to them in the 1960s and 1970s. He used educational level as an indirect measure of intelligence and showed that better-educated parents had fewer children than those less well educated. Further evidence for dysgenic fertility for intelligence in the United States has been obtained by Osborne, Vining, and Retherford. Vining has also found some evidence for dysgenic fertility in Sweden, as has Retherford in Japan. Lynn has shown that dysgenic fertility for both intelligence and conscientiousness has been present throughout the world in the 20th century and is greater in the economically developing countries than in the industrialized nations.

In the 1980s another intellectual conflict emerged, growing out of race differences in IQ and cognitive test performance. Affirmative action policies were established that favored less qualified blacks in admission to universities and in hiring by industry. These policies have been opposed on libertarian grounds and also on efficiency grounds, as IQ tests are among the best predictors of educational success and job performance. Van den Haag has discussed the ethical contradictions and drawbacks of such policies, as has Levin. Gottfredson, Humphreys, and Levin have examined their practical economic consequences.

Several Pioneer scholars have written about the educational problems arising from the average differences in IQ of different racial and ethnic groups. The case against forced integration and busing was argued by van den Haag and Scott on libertarian and efficient teaching grounds. Hardin has examined both the demographic and the environmental consequences of unrestricted immigration from over-populated third world nations.

THE FIRST 60 YEARS...

In the 60 plus years of its existence, the Pioneer Fund has had some considerable successes in the research programs it has funded on the nature of human nature. In 1937, the year in which the Fund was established, virtually nothing was known about the extent to which intelligence and personality are determined by inheritance. By the end of the twentieth century a firm foundation of scientific knowledge had been laid. A consensus has been reached among the academic community that genetic factors play a substantial role in the determination of both intelligence and personality. The extent to which genetic factors play a part in race differences in intelligence and personality remains controversial. There is, however, accumulating evidence that this is the case.

Less progress has been made, however, in conveying these conclusions to the media and the informed public. This has been largely due to the success of a small group of ideologically-motivated academics and political activists. They have convinced large segments of the popular media, as opposed to the professional journals and forums, that the equalitarian dogma is solidly supported by empirical research and that the hereditarian research program is inherently ethically suspect. This mismatch between expert opinion and mass media coverage is best shown by the results of the Snyderman and Rothman study.

In their 1988 book, *The IQ Controversy: The Media and Public Policy*,[2] Mark Snyderman and Stanley Rothman examined the relationship between the research in the social sciences, media coverage, and public policy. They surveyed experts in behavioral genetics and psychometrics and also examined how these issues were reported in the newspapers and on television. Snyderman and Rothman found that, despite frequent media misrepresentation to the contrary, the results of the research funded by Pioneer are in fact at the core of the expert scientific consensus.

...AND THE NEXT

Much work still remains to be done. First, although a great deal of evidence has accumulated on the importance of heredity in the determination of intelligence and personality, and on race differences, this needs strengthening by further research, especially at the gene level, which is now possible. Second, the importance of heredity in human affairs needs to be conveyed to opinion makers in the media, to politicians, and to the general public. While scientific facts do not yield ethical or policy conclusions, they are vital in determining the extent to which desired policies are pragmatically possible. The final and most important task, therefore, is to disseminate the results of over 60 years of Pioneer Fund research to the general public. Informed discussion of the implications of this body of scientific evidence for public policy ultimately rests with the citizens.

These are the tasks that lie ahead for the Pioneer Fund, the scientific community, and society in the 21st century. Hopefully, this book is the start of a new beginning, the first project of the next 60 years.

NOTES

1. Browne, M. 1994. What is Intelligence and Who Has It? A Review of *The Bell Curve* by Richard J. Herrnstein & Charles Murray, *Race, Evolution, and Behavior* by J. P. Rushton, and *The Decline of Intelligence in America* by Seymour W. Itzkoff. *New York Times Book Review*. October 16.

2. Snyderman, M. & Rothman, S. 1990. (2nd Paperback Printing). *The IQ Controversy: The Media and Public Policy*. New Brunswick, NJ: Transaction.

Appendix

Certificate of Incorporation of the Pioneer Fund, Inc.

The Pioneer Fund was incorporated on 11 March 1937. The certificate of incorporation stated that the Fund had two objectives: First, to provide financial assistance to the parents of children likely to become socially valuable citizens who would make important contributions to their society; and second, to provide grants for research into the study of human nature, heredity, and eugenics. Only the second objective has been funded.

Filed: March 17, 1937
Amended: June 14, 1985

CERTIFICATE OF INCORPORATION

of

THE PIONEER FUND, INC.

Pursuant to the Membership Corporations Law

WE, THE UNDERSIGNED, for the purpose of forming a membership corporation pursuant to the Membership Corporations Law of the State of New York, hereby certify:

1. The name of the proposed corporation is THE PIONEER FUND, INC.

2. The purposes for which it is to be formed are:

To acquire money, securities, or other property, real or personal, by gift, legacy, or otherwise, including the right to receive the income or principal of any property, legacy or devise given by will or otherwise in trust to pay the principal or income to this Corporation; and to hold, invest, use, and dispose of the principal and income of the same for any one or more of the following charitable purposes:

A. To provide or aid in providing for the education of children of parents deemed to have such qualities and traits of character as to make such parents of unusual value as citizens, and in the case of children of such parents whose means are inadequate therefor, to provide financial aid for the support, training, and start in life of such children.

The children selected for such aid shall be children of parents who are citizens of the United States, and in selecting such children, unless the directors deem it inadvisable, consideration shall be especially given to children who are deemed to be descended predominantly from persons who settled in the original thirteen states prior to the adoption of the Constitution of the United States and/or from related stocks, or to classes of children the majority of whom are deemed to be so descended.

Subject to the requirement that the Corporation shall be administered for strictly charitable objects, and in so far as it may be found practicable so to do, the foregoing purposes shall be carried out in such manner as to give assurance to parents of the character described that their children shall not lack an adequate education or start in life and thus to encourage an increase in the number of children of such parents, and in so far as the qualities and traits of such parents are inherited, to aid in improving the character of the people of the United States.

B. To conduct or aid in conducting study and research into the problems of heredity and eugenics in the human race generally and such study and such research in respect to animals and plants as may throw light upon heredity in man, and to conduct or aid in conducting research and study into the problems of human race betterment with special reference to the people of the United States, and for the advance of knowledge and the dissemination of information with respect to any studies so made or in general with respect to heredity and eugenics.

The Corporation is not organized for pecuniary profit and shall not engage in any activities for pecuniary profit, and no officer, director, member, or employee of the Corporation shall receive any pecuniary profit from the operations thereof except reasonable compensation for services in effecting or carrying out one or more of its activities or as a proper beneficiary of its strictly charitable purposes. Any and all property acquired by the Corporation shall be held, used, and disposed of for charitable purposes only and the above stated specific purposes shall be interpreted in a manner consistent with this intention.

3. Its operations are to be conducted principally in the territory comprising the continental United States, including the District of Columbia.

4. Its office to be located in the City, County and State of New York.

5. The number of its directors shall be not less than 3 nor more than 9.

6. The names and residences of the directors until the first annual meeting are:

Names	Residences
Wickliffe Preston Draper	322 East 57th St. New York City, N.Y.
Harry H. Laughlin	Cold Spring Harbor, Long Island, N.Y.
Malcolm Donald	638 Blue Hill Ave. Milton, Mass.
Frederick Henry Osborn	Garrison-on-Hudson, New York.

7. All of the subscribers to this certificate are of full age; at least two-thirds of them are citizens of the United States; at least one of them is a resident of the State of New York. Of the persons named as directors, at least one is a citizen of the United States and a resident of the State of New York.

The Secretary of State is designated as agent of the Corporation upon whom process against the Corporation may be served. The post office address within or without the state to which the Secretary shall mail a copy of any process against the Corporation served upon him is 299 Park Avenue, 17th Floor, New York, New York 10171.

IN WITNESS WHEREOF, we have made, subscribed, and acknowledged this certificate as of this 27th day of February, 1937.

Wickliffe Preston Draper

Harry H. Laughlin

Malcolm Donald

Frederick Henry Osborn

Vincent R. Smalley

Name Index

Subject Index